The Smart Swarm

The Smart Swarm

How Understanding Flocks, Schools, and Colonies Can Make Us Better at Communicating, Decision Making, and Getting Things Done

PETER MILLER

AVERY
a member of Penguin Group (USA) Inc.
New York

Published by the Penguin Group
Penguin Group (USA) Inc., 375 Hudson Street, New York, New York 10014, USA • Penguin Group (Canada), 90 Eglinton Avenue East, Suite 700, Toronto, Ontario M4P 2Y3, Canada (a division of Pearson Penguin Canada Inc.) • Penguin Books Ltd, 80 Strand, London WC2R 0RL, England • Penguin Ireland, 25 St Stephen's Green, Dublin 2, Ireland (a division of Penguin Books Ltd) • Penguin Group (Australia), 250 Camberwell Road, Camberwell, Victoria 3124, Australia (a division of Pearson Australia Group Pty Ltd) • Penguin Books India Pvt Ltd, 11 Community Centre, Panchsheel Park, New Delhi–110 017, India • Penguin Group (NZ), 67 Apollo Drive, Rosedale, North Shore 0632, New Zealand (a division of Pearson New Zealand Ltd) • Penguin Books (South Africa) (Pty) Ltd, 24 Sturdee Avenue, Rosebank, Johannesburg 2196, South Africa

Penguin Books Ltd, Registered Offices:
80 Strand, London WC2R 0RL, England

Published simultaneously in Canada

Library of Congress Cataloging-in-Publication Data

Miller, Peter.
The smart swarm : how understanding flocks, schools, and colonies can make us better at communicating, decision making, and getting things done / Peter Miller.
p. cm.
Includes bibliographical references and index.
ISBN 978-1-58333-390-7
1. Animal behavior. 2. Human behavior. 3. Decision making. I. Title.
BF671.M555 2010 2009048619
156—dc22

Printed in the United States of America
1 3 5 7 9 10 8 6 4 2

BOOK DESIGN BY AMANDA DEWEY
ILLUSTRATIONS BY OLIVER UBERTI

For my wife, Priscilla,
and my parents, Mary Lou and Bob

Contents

Foreword by Don Tapscott ix

Introduction
When in Doubt, Turn to the Experts xiii

· 1 ·

Ants
Who's in Charge Here? 1

· 2 ·

Honeybees
Making Smart Decisions 33

· 3 ·

Termites

One Thing Leads to Another 105

· 4 ·

Birds of a Feather

Secrets of Flocks, Schools, and Herds 159

· 5 ·

Locusts

The Dark Sides of Crowds 227

Conclusion

Doing the Right Thing 259

Acknowledgments 270

Notes 274

Index 279

Foreword

BY DON TAPSCOTT

Many people have studied how crowds, mass collaboration, business ecosystems, and networks transform the way organizations can do things better. I, for one, am convinced that we're in the early days of the biggest change to the deep structures, architecture, and modus operandi of the century. But it often feels a lot more like an art than a science.

It turns out that nature can help us with the science.

When it comes to organizing ourselves in society, we often default to traditional hierarchies. This model has worked well as a way of systematizing work, establishing authority, deploying resources, allocating tasks, defining relationships, and enabling organizations to operate. Whether the ancient slave empires of Greece, Rome, China, and the Americas; the feudal kingdoms that later covered the planet; the corporations of industrial capitalism; or the bureaucracies of Soviet-style communism, hierarchies have been with us since the dawn of human history. Even the management literature today that advocates empowerment, teams, and networking takes the command-and-control method as a premise: Every person in an organization is subordinate to someone else. Hierarchies also define the relationships among companies. Every company is positioned in a supply chain whose subordinate companies it controls, and it is in turn beholden to the clients or customers it serves. In the old model of economic development, worker bees are to be supervised in their honey production.

The basic concept is here to stay, but traditional hierarchies have increasing limitations. More than twenty years ago, Peter Drucker described managers as "relays—human boosters for the faint, unfocused signals that pass for information in the traditional, pre-information organization." Communication from the bottom up is often limited, except through formal labor-management relations. Hierarchies are typically bureaucratic, and employees lack motivation. Increasingly, they are insufficient as a way of organizing for the fast-paced economy where human capital needs to be unleashed for innovation, value creation, and customer relationships.

Then along comes the Internet, a communication medium that radically drops transaction and collaboration costs. This changes two very fundamental things about the protocol of enterprises. First, there are alternatives for organizing the internal workings of companies and other organizations. As Peter Miller describes, companies like Best Buy can harness the wisdom of many with techniques such as prediction markets to operate more effectively, and in doing so they challenge basic tenets of hierarchical control. Peers can collaborate across organizational silos. We can rethink power, now achievable *through* people rather than over people. Work can be organized on new project models, where the genius of human capital can be unleashed from its old command-and-control constraints. Employees can forge their own self-organized interconnections and form cross-functional teams capable of interacting as a global, real-time work force. Loosening organizational hierarchies and giving more power to employees can lead to faster innovation, lower cost structures, greater agility, improved responsiveness to customers, and more authenticity and respect in the marketplace.

Second, the boundaries of firms can become more porous, enabling powerful new approaches for orchestrating ability to innovate, to create goods and services, and even to produce public value. Rather than hierarchical supply chains, firms can build peer-to-peer networks where the roles, motivations, and behavior of the players are different—with dramatically better results.

What's missing is a better science to all this, which is where *The Smart Swarm* comes in. What could we learn from the dynamic, complex "collaborations" that exist in nature itself? What can nature tell us to help us bring complexity theory down to earth?

In the past, I and others have compared the networked organizations springing up to a skein of geese forming a V—acting in unison but without centralized control. Some years ago, Thomas Stewart, former editor of the *Harvard Business Review*, explained that the motion of the group is the aggregate result of the actions of each individual animal acting on the basis of its local perception of the world. There is no one leader. The bird at the front of the V has to work hardest because of wind resistance. But when it gets tired, another bird takes the leadership position. The birds have a collaborative leadership of sorts.

With the publication of *The Smart Swarm*, for the first time, the lessons of flocks, schools, and colonies have been brought together in a readable text about how to get things done better. In a sense it's a step toward creating a science of collaboration.

And where is all this going? Is it possible that as everyone connects through the global digital platform, we can begin to share not only information but also our ability to remember, process information, and even think? Is this just a fanciful analogy, or will we come to consider networking as the neural routes that are growing to connect human capital and transforming, again, quality (of connections) into quantity (something fundamentally new)?

You'll enjoy this book, and not only for its speculation about the future. Rather, it's full of practical guidelines about what nature can tell us about how to build better organizations today. How does your company embrace self-organization, diversity, knowledge, individual collaboration, and adaptive mimicking to outdo your competitors or deliver better value to society. And how can you avoid the dark side of smart swarming?

Read on.

Don Tapscott is the author of fourteen books about new technology in business and society, most recently (with Anthony D. Williams) MacroWikinomics: Rebooting Business and the World *(September 2010).*

Introduction

When in Doubt, Turn to the Experts

Not long ago Southwest Airlines was wrestling with a difficult question: Should it abandon its long-standing policy of open seating on planes? Of all the major airlines, Southwest was the only one that let passengers choose where to sit once they got on board. The airline had done it that way for more than thirty-four years, and it took pride in being an industry maverick. The company's independent attitude had helped make it one of the largest airlines in the world. Southwest, remember, was the first carrier to encourage flight attendants to tell jokes in the air.

Lately, though, some customers, especially business travelers, had complained that the free-for-all to get on a Southwest plane was no fun. To obtain a good seat, travelers had to arrive at the airport hours before their flight to secure a place at the head of the line, or remember to print out a boarding document the day before from the company's online reservation system. Some said the process made them feel more like cattle than customers, which, in the competitive airline business, was a problem. So Southwest put the issue on the table; if assigned seating would make people happier, the company was willing to consider it.

The question turned out to be more complicated than it seemed. For one thing, no one knew how assigned seating would affect the amount of time it would take for Southwest to board everybody. If assigned seating made the process faster, then switching made sense, of course. But if it slowed things down, it wouldn't help. Boarding speed depended, in part, on which pattern was used. Should the company start in the back of the plane and work forward? Should it start in the front and move to the back? What about boarding window seats first, then middle seats, and then, finally, aisle seats? How about alternating among various zones? Each strategy offered advantages and disadvantages, and each required a different amount of time. Given such variables, how was the airline supposed to make a decision?

To a Southwest analyst named Doug Lawson, the answer seemed obvious: the best way to determine whether assigned seating would be faster was to create a computer simulation of passengers boarding a plane, and then try out one pattern after the other. Other airlines had done more or less the same thing over the years. But Lawson's plan had a difference—it was based on the behavior of ants.

"Ants were a good fit for this study, because we had all these individuals pouring into a tight space, interacting with one another," he says. "Every individual had a task to do—in this case, obtain a seat—while dealing with all the others doing the same thing. In a way, it was a typical biological problem."

Like real ants, Lawson's digital ones followed a few simple rules to guide their behavior. "Each ant was allowed to go down the jet ramp and wander onto the plane. If we were simulating open seating in that run, each ant had its own idea of a good place to sit, based on actual passenger data, and it would look over the situation and say, well, I see that seat is open. I'm going to try to get over to that one." If the path was clear, then the ant moved down the aisle to the appropriate row and took its seat. If the path was blocked by other ants, it either waited a few seconds, or asked them to move aside. (Lawson had to add the waiting rule after a few raucous simulations. "We had all these ants trying to get in through the galley, and they were pushing and shoving and bouncing off each other," he says. "They were creating chaos on the plane, so we had to tone some of them down.")

As soon as all the ants were seated, the simulation was finished, and its elapsed time could be compared with those from other runs. Since Southwest flies only Boeing 737s, the physical constraints of the problem were always the same, which made it easier to calibrate Lawson's simulations with data from actual boardings. In addition, Southwest staged a full day of experiments using employees on a real plane to ground-truth the results. What Lawson determined from all this, after repeating his simulations for every feasible pattern, was that open seating was relatively fast, but that assigned seating, under certain circumstances, could be faster. The difference, though, was only a minute or two—not enough, by itself, to abandon Southwest's long-standing tradition.

"We have a lot of loyal customers who just like walking onto the plane and sitting with whomever they want to," Lawson observes. "They saw that as part of our brand, and they didn't want the brand changed at all."

So instead of dumping open seating, the airline took another close look at the way passengers were lined up at the gate. If the real problem was that people didn't like competing for a spot in line, Southwest figured, then why not assign them a spot when they checked in, so they wouldn't have to worry about it later? Boarding would still be first-come, first-served, but passengers' places in line would be held as soon as they checked in, whether in person or online. That way, passengers wouldn't have to show up hours ahead of time and hold their places, and when they got on the plane they could still sit anywhere they wanted, "as long as they didn't sit on top of somebody else," Lawson says. Southwest adopted this new system in late 2007.

Why was an ant-based simulation a good idea for Southwest? What do ants and airlines have in common? The answer has to do with the remarkable phenomenon I call a *smart swarm*. Evolved over millions of years, a smart swarm might be a colony of ants in the desert that has figured out exactly how many workers to assign to various jobs each morning, despite an unpredictable environment. It might be a hive of honeybees in the forest that has worked out a foolproof system to choose just the right tree for a new home, despite conflicting opinions among many individuals. It might be a school of thousands of fish in the

Caribbean Sea that knows how to coordinate its behavior so precisely that it can change direction in a flash, as if it were a single silvery creature. Or it might be a vast herd of caribou on an epic migration to an Arctic coastal plain, each animal certain of reaching the calving grounds even though most haven't got a clue about exactly where they're going. Simply put, a smart swarm is a group of individuals who respond to one another and to their environment in ways that give them the power, as a group, to cope with uncertainty, complexity, and change.

Inspired by the practical way in which an ant colony splits a big problem into thousands of little ones, for example, Lawson set out to tap into the same kind of swarm intelligence with virtual ants he called "cognitive moving objects." Although his digital insects were highly simplified simulations, they were designed to capture the fundamental cleverness of real ant colonies. "Down here in Texas we have lots of different types of ants," says Lawson, who works at Southwest's headquarters in Dallas. "Take the leaf-cutting ants of central Texas. They have the most amazing social structure you could imagine." Like their tropical cousins in South America, this ant species (*Atta texana*) employs an assembly line of workers to farm a symbiotic fungus, which the colony eats. At one end of the assembly line, skillful workers cut pieces of leaves from trees or bushes and carry them back to the nest, as biologists Bert Hölldobler and E. O. Wilson describe in their book *The Superorganism*. Inside the nest, a second group of workers, slightly smaller in size than the first, snips the leaves into tiny pieces and leaves them for the next group. The third group of even smaller workers chews the pieces into pulp and shapes the pulp into pellets. Then a fourth group of still smaller workers plants strands of fungus inside a pile of pellets in the colony's subterranean garden. Finally, the smallest workers of all lovingly tend the fungi, removing unwanted spores. "That's how that shop is run," Wilson says.

With several million workers per nest, a leaf-cutter colony can harvest a half-ton or so of vegetation a year, which gives you some indication of the incredible power that ants acquire by combining and coordinating their efforts. Such abilities, managed through a sophisticated communication system based on chemicals, has enabled ant colonies to raise their behavior as a group to a level far above that of individual ants. Which is why Wilson and Hölldobler describe such colonies as superorganisms. "The modern insect societies," they write, "have a vast amount to teach us today."

To some people, this may come as a surprise. How could ants, bees, or termites know something we don't? How could such insignificant creatures solve complicated problems better than an $11-billion-a-year airline like Southwest? If ants are so smart, why aren't they flying the 737s? The fact is, these and other creatures have been dealing with the most difficult kinds of problems for millions of years: Will there be enough food for the colony this week? Where will it be found? How many workers will the hive need to build a nest? How will the weather affect the herd's migration this year? The way they've responded to these challenges has been to evolve a special form of group behavior that is flexible, adaptive, and reliable.

Translated into mathematical formulas, the principles of a

smart swarm have given businesses powerful tools to untangle some of the knottiest problems they face. Manufacturing companies have experimented with them to optimize production, for example. Telephone companies have tested them to speed up calls. Aircraft mechanics and engineers have applied them to identify problems in new airplanes. And intelligence agents have used them to keep track of a dangerous world.

How does a smart swarm work? We'll find out in the first three chapters by following biologists into the field to unlock the secrets of collective behavior. As these researchers have discovered, social insects such as ants, bees, and termites distribute problem solving among many individuals, each of which is following simple instructions but none of which sees the big picture. Nobody's in charge. Nobody's telling anybody else what to do. Instead, individuals in such groups interact with one another in countless ways until a pattern emerges—a tipping point of motion or meaning—that enables a colony of ants to find the nearest pile of seeds, or a school of herring to dodge a hungry seal.

In the fourth chapter we'll look at the subtle role that individuals play in keeping a group on course. For groups such as flocks of birds, schools of fish, or herds of caribou, which are made up of individuals largely unrelated to one another, the key to survival demands a set of skills that balances group behavior with self-interest. As humans, we share many common problems with such groups, since we're often torn by the same impulses—to cooperate but also to profit, to do what's right for the community but also to look out for ourselves and our families.

Not every swarm is smart, of course. Group behavior also has a dark side. In the fifth chapter we'll find out what scientists have discovered about locusts to explain why peaceful groups of grasshoppers suddenly explode into voracious plagues. To learn how human instincts can go haywire, we'll also follow the work of researchers who have studied fatal crowd disasters among religious pilgrims in Saudi Arabia—and what's been done to prevent such accidents in the future. What separates a smart swarm from its stupid cousin? Why does a happy crowd suddenly turn into a rampaging mob? The reason, simply put, is that a smart swarm uses its collective power to sort through countless possible solutions while the mob unleashes its chaotic energy against itself. And that makes it so important to understand how a smart swarm works—and how to harness its power.

As everyday life grows more complicated, we increasingly find ourselves facing the same problems of uncertainty, complexity, and change, drowning in too much information, bombarded with too much instant feedback, facing too many interconnected decisions. Whether we realize it or not, we too are caught up in worlds of collective phenomena that make it more difficult than ever to guide our companies, communities, and families with confidence. These challenges are already upon us, so we need to be prepared. The best way to do that, as you'll see in the pages ahead, is to turn to the experts—not the ones on cable TV but those in the grass, in the air, in the lakes, and in the woods.

1

Ants

Who's in Charge Here?

Just off Route 533 in southwestern New Mexico, a barbed-wire fence surrounds sixty acres of what used to be a sprawling cattle ranch at the foot of the Chiricahua Mountains. Some years ago, at the request of biologist Deborah Gordon, Stanford University bought the property to keep out of the hands of developers a small research site she'd established. But the subdivisions and convenience stores never materialized. In fact, not much at all has happened on this little patch of the Sonoran Desert to disturb the current residents of the site, including several hundred colonies of red harvester ants (*Pogonomyrmex barbatus*). For more than two decades now, Gordon has documented the life histories of these colonies, where, day in and day out, season after season, ants go about their business with a curious mix of efficiency and utter chaos.

The workday starts early at Colony 550, an older nest of some ten thousand ants near the eastern border of the site. From dawn to midmorning, one group after another emerges from the nest to carry out various tasks. The first on the job are patrollers, who poke their heads out of the entrance hole just before sunrise. Appearing to be in no hurry, they mill around on the circular nest mound, inspecting the pebbly surface like groundskeepers at a golf course assessing the health of a green. If something has happened during the night, patrollers will be the first ants to know. Has the rain left a pile of debris on a foraging trail? Has the wind redistributed the seeds the ants collect for food? What are the neighbors up to this morning? As they wander farther and farther from the nest entrance, patrollers may bump into scouts from nearby colonies doing exactly the same thing, and, if they do, forager ants from both sides might later fight. "Last week, for some reason, we noticed quite a few foragers walking around with the heads of other ants attached to their bodies," says Mike Greene, a biologist from the University of Colorado–Denver who was doing research at the site. "They'd clearly been having little ant wars."

The patrollers are soon joined by a crew of nest maintenance workers, each carrying a bit of dirt, seed husk, or other trash up from below ground. In contrast to the patrollers, they seem narrowly focused on their tasks, searching for a suitable place to deposit their loads. The moment they find one, they drop what they're carrying, turn around, and head back down into the nest.

Next come a handful of midden workers, who tidy up what the maintenance workers have left behind. Not that they do this in any sensible way. If you watch one working for a while, Greene says, you'll probably find it puzzling. "Midden workers remind me of my fifteen-month-old daughter. They take an object from Point A and drop it at Point B. Then they pick something else up and go to Point C. It all seems very random." A time-lapse movie of the morning's activity, though, would show a pile of dirt and ant trash steadily growing along one edge of the nest mound. "So it turns out they're organized, after all," he says.

The last to appear are the foragers, who greatly outnumber the other workers. Streaming out of the entrance hole, they charge directly for the tall grass that rings the nest mound and disappear into a sea of Mormon tea, acacia, and snakeweed. Following ant highways through the underbrush, the foragers may venture as far as sixty feet from the nest in search of seeds. Because these seeds, for the most part, have ridden the winds from other parts of the desert, rather than coming from plants on the site, they tend to be scattered in unpredictable ways. So it could take a forager as long as twenty minutes to find one. As soon as it does, it picks up the seed and carries it straight back to the nest.

By nine a.m., the nest hole has taken on the appearance of a frantic subway entrance, with ants rushing in and out. In a colony like 550, which is nearly twenty years old, the nest may be six feet deep. Down below, in an elaborate network of tunnels and chambers, as Gordon describes in her book *Ants at Work,* other groups of ants are busily stacking seeds in storage chambers, according to size and shape; removing dead ants, grasshopper legs, and other unwanted objects from the nest; tending brood; caring for the queen; or simply standing ready in reserve.

From top to bottom, Colony 550 seems to be a model of efficiency, with each group performing its task in an orderly sequence—an impression strengthened by each ant's habit of constantly touching its antennae with every other ant it meets, as if to make sure that everybody's on the same page. From patrollers and maintenance workers to midden workers and foragers, every member of the colony seems to be following a master plan, like tiny cogs in a machine or the employees of a successful factory.

But that's not what's happening here at all.

Despite its well-managed appearance, Colony 550 does not function like any organization you are ever likely to encounter. It has no bosses, managers, or supervisors of any kind. The queen, despite her lofty title, wields no authority. Her sole function is to lay eggs, not to give commands. When patrollers venture out into the grass, they're not taking orders from a squad leader. When nest maintenance workers repair a tunnel, they're not following any blueprints. Young ants entering the work force don't have to sit through an orientation meeting or memorize a mission statement, because they never need to see the big picture. No ant ever understands the purpose of its own labor, why it needs to complete the job, or how it fits in.

Yet the colony does just fine. Consider the way it responds quickly and effectively to changes in its environment. If patrollers this morning discover a tasty pile of seeds, additional ants will head out to look for more within minutes, and these additional ants will become foragers. Did last night's storm damage the nest? More maintenance workers will show up to repair it, even if that requires younger nurse ants to pitch in. Depending on the challenge or the opportunity, the colony as a whole calculates quickly and precisely how many workers are needed to take care of a job, then adjusts its resources accordingly.

This flexible system, evolved during 140 million years of ant history, is one of the main reasons that the world's fourteen thousand or so known species of ants have flourished in a bewildering variety of ecosystems, from tropical rain forests to city sidewalks. Their way of doing things may look messy, but it enables them to accomplish amazing feats, such as organize highways, build elaborate nests, and stage epic raids—all without any leadership, game plan, or the least sense of mission.

How do they do it?

Ants Aren't Smart

Every morning in August, Deborah Gordon sets out from the Southwestern Research Station near Portal, Arizona, and drives just across the border into New Mexico to observe red harvester ants. Every afternoon, once the ants have retreated underground to escape from the blazing heat, the biologist returns to the station with a renewed sense of wonder—not that the ants are so skillful at what they do, but that they appear to be such little dummies.

"If you watch an ant try to do something, you'll be impressed by how inept it is," she says. "Often, it doesn't go about things the way you think would be best, it doesn't remember anything for very long, and it doesn't seem to care if it succeeds." Only one in five ants actually accomplishes what it sets out to do. "The longer you watch an ant the more you end up wanting to help it."

Gordon doesn't study ants as individuals, though. Her research focuses on the behavior of ant *colonies*. As colonies, she says, ants are capable of solving problems far beyond the abilities of individuals, such as how to find food, allocate resources, or respond to competition from neighbors.

"*Ants* aren't smart," she clarifies. "Ant *colonies* are."

The central focus of Gordon's research has been the ants' system of task allocation, which is how a colony decides which jobs need to be done on any particular day. Given all the uncertainties that red harvesters face—from the iffy availability of food to competition from neighbors—a colony must calculate as a group how many workers to send out foraging, how many to keep on patrol, how many to hold back to tend brood, and so on.

"One of my favorite moments in the movie *Antz* is a scene I call the Bureau of Task Allocation," she says of the 1998 DreamWorks animated film. "The ants are brought to some bureaucrats—they've got clipboards—behind a counter, and each ant is just stamped, and given its task. This, of course, is the way we organize our work, where certain individuals have the job of assigning work to other individuals. So it's easy for us to imagine that there's somebody in there with a clipboard, telling somebody else what to do." But that's not how the ants do it.

To understand the real process of task allocation, Gordon and fellow biologist Mike Greene conducted a series of experiments a few years ago with foragers. They knew that a colony, depending on circumstances, doesn't forage every day. It might be too cold or windy to go outside, or there might be a hungry lizard waiting at the edge of the nest mound. Patrollers seem to be the key to this decision. As they return from their early-morning scouts of the neighborhood, they're greeted near the nest entrance by a crowd of foragers. The foragers touch antennae with the patrollers, and if they bump into the right number of patrollers, the foragers are more inclined to go out. The behavior of the patrollers, in other words, informs the decisions of the foragers.

It doesn't happen in the way you might expect, though. "The patrollers aren't passing along anything elaborate," Gordon says. "They're not coming back and giving instructions to the foragers, saying go here and do this. The message is merely in the contact. And that's what's hardest for us to understand, because we keep falling into the temptation to think that they're doing it the way that we would."

To get to the bottom of this group-oriented behavior, she and Greene conducted an experiment using fake patrollers. First they captured real patrollers leaving several colonies one morning. Then, after waiting thirty minutes, they dropped tiny glass beads coated with the smell of patrollers into each nest entrance. Red harvesters, like most ants, are covered with a layer of grease that keeps them from drying out. This grease, made of hydrocarbons, carries an odor specific not only to their colony but also to their task group. "For the ants, you might say, chemicals are what vision is for us," Greene says. When foragers inside the nest encountered the glass beads coated with patroller hydrocarbons, they took them for real patrollers.

What Gordon and Greene wanted to know was whether the *rate* at which foragers encountered patrollers made any difference. If it did, that might represent an important mechanism in the colony's decision-making process. So they varied the speed at which they dropped patroller beads into each nest. In the first of four trials, they added one bead every three minutes. In the second, one bead every forty-five seconds. In the third, one bead every ten seconds. In the last, one bead every second. The results were dramatic.

In the first two trials, the relatively slow rates prompted few foragers to go out. The same was true of the fourth trial with the fastest rate. But in the third trial, when foragers encountered glass beads at just the right rate—one bead every ten seconds—they left the nest in a big rush with four times as many foragers.

"The rate needs to be about ten seconds because that must be how long an ant can remember what happened to it," Gordon says. "If an ant has to wait forty-five seconds to meet another ant, it forgets the previous one. It's as if the encounter never happened." Red harvesters, it seems, have a very short attention span. If the rate is too fast, meanwhile, that may mean that something has driven foragers back to the nest, such as a predator. The rate has to be just right.

A forager's decision, that is, doesn't depend on it receiving instructions from a patroller or figuring out on its own what's needed. It depends instead on the ants following a simple rule of thumb: If it meets the right number of patrollers returning at the right rate, it goes out looking for seeds. If it doesn't, it stays put. "Nobody's deciding whether it's a good day or not to forage," Gordon says. "The collective is, but no particular ant is."

Once the first foragers leave the nest, a separate mechanism kicks in to regulate the total number of foragers that go out that day. The key encounters this time take place between foragers only. As successful foragers return to the nest with seeds, they're met at the nest entrance by foragers waiting in reserve. This contact stimulates the inactive ants to go out. Foragers normally don't come back until they find something. So the faster the foragers return, the faster other ants go out, enabling the colony to tune its work force to the probability of finding food.

This simple rule, applied by one forager after another in the crowded space near the entrance hole, functions like a simple calculator for the colony. The sum of all the decisions by all the ants gives the colony the answer to the question "How many foragers do we need searching for food today?"

The ants aren't smart. The colony is.

THIS INTRIGUING BEHAVIOR, of course, isn't unique to ants. Many groups of animals, from honeybees to herring, tackle difficult problems without direction from leaders. They do it through a phenomenon that scientists call *self-organization*—the first principle of a smart swarm. Although examples of self-organization appear all around us in nature, scientists have studied it intensively only during the past few decades. First described by chemists and physicists, the term originally referred to the spontaneous appearance of patterns in physical systems, such as the rippling of sand dunes or the hypnotic spirals that form when certain chemical reactants are combined. Later it was adopted by biologists to explain the intricate structure of wasp nests, the synchronized flashing of some species of fireflies, and the way that swarms of bees, flocks of birds, and schools of fish instinctively coordinate their actions.

What these phenomena all have in common is that none of them is imposed from the top by a master plan. The patterns, shapes, and behaviors we see in such systems don't come from preexisting blueprints or designs, but emerge on their own, from the bottom up, as a result of interactions among their many parts. We call an ant colony self-organizing because nobody's in charge, nobody knows what needs to be done, and nobody tells anybody else what to do. Each ant goes through its day responding to whatever happens to it, to the other ants it bumps into, and to changes in the environment—what scientists call "local" knowledge. When an ant does something, it affects other ants, and what they do affects still others, and that impact ripples through the colony. "No ant understands its own decisions," Gordon says. "But each ant's decision is linked to another ant's decision and the whole colony changes."

Although the ultimate origins of self-organization remain something of a mystery, researchers have identified three basic mechanisms by which it works: *decentralized control*, *distributed problem-solving*, and *multiple interactions*. Taken together, these mechanisms explain how the members of a group, without being told to, can transform simple rules of thumb into meaningful patterns of collective behavior.

To get a feel for how these mechanisms work, consider a day at the beach with your family or friends. When you first arrive, you don't stand around waiting for someone to give you instructions. Apart from certain restrictions imposed by the community (no nudity, no pets, no alcohol, for example) you're on your own. Nobody tells you where to sit, what to do, whether to go into the water or not (unless the lifeguard gets bossy). Everybody can do pretty much what they want to, which is one way of describing *decentralized control.*

If it's a beautiful day and the beach is crowded, of course, it might take some time to find the perfect place to sit down. You don't want to choose a spot too close to the water, or your beach chairs and blanket could get soaked by a big wave. Nor do you want to sit far away from the water, where you can't feel the ocean breeze. If you plan to go swimming, it might be convenient to choose a location near the lifeguard, as every family with little children has already figured out (which is why all those umbrellas are clustered around the guard's stand). In the end, you choose a space with just enough room to spread your blanket yet maintain the proper distance in all directions from your neighbors' blankets, which is the unspoken rule of thumb at the beach. If you could look down from a helicopter, you'd see a mosaic of blankets evenly spaced from one another, reflecting the success of the crowd's *distributed problem-solving.*

Then something curious happens. Just as you're settling into your beach chair with Stephen King's latest novel, you notice that a few people have stood up to look at the water. Then a few more do the same thing. And a few more. Suddenly it seems like everybody's standing and looking at the water, so you do too. You don't have any idea why, but you're suddenly alert, full of questions. What's going on? Is somebody drowning? Is there a shark? What's everybody looking at? What began, perhaps, as a simple act of curiosity by a few individuals—staring at the water—spreads from person to person down the beach, snowballing into a collective state of alarm. That's how infectious *multiple interactions* can be. And the impressive thing is, if there *had* been a shark, everybody would have found out about it almost as quickly as if someone had shouted "Jaws" with a bullhorn.

"If we each respond to little pieces of information, and we follow certain rules, the whole crowd will organize in a certain way," Mike Greene says, "just like when we're looking down on an ant colony, we can actually see its behavior change, even though none of the ants is aware of it."

Day in and day out, that is, self-organization provides an ant colony like 550 with a reliable way to manage an unpredictable environment. Wouldn't it be useful if we could do the same thing?

The Traveling Salesman Problem

One afternoon in the summer of 1990, an Italian graduate student named Marco Dorigo was attending a workshop at the German National Research Center for Computer Science near Bonn. At the time, Dorigo was working on a doctoral thesis in Milan about ways to solve difficult computational problems. The talk he'd come to hear was by Jean-Louis Deneubourg, a professor from the Free University of Brussels, about his research with ants. "I was already interested in ways that natural systems could be used as inspiration for information science," Dorigo says. "But this was the first time anybody had made a connection between ant behavior and computer science."

In his presentation, Deneubourg described a series of experiments that he and his colleagues had done with common black ants known as Argentine ants (*Iridomyrmex humilis*). Like many ants, this species leaves a trail of chemical secretions when foraging. Such chemicals, called pheromones, come from glands near the tip of the ant's abdomen, and they act as powerful signals, telling other ants to follow their trails. Foragers normally lay down such trails after they have found a promising source of food. As they return to the nest, they mark their paths so that other ants can retrace them to the food. But Argentine ants are different. They lay down pheromone trails during the search phase as well. That appealed to Deneubourg, who was curious about how foragers decided where to explore.

In one experiment in his lab, Deneubourg and his colleagues placed a bridge between a large tub containing a colony of Argentine ants and another tub containing food. The bridge had a special design. About a fourth of the way across, it split into two branches, both of which led to the food, but one of which was twice as long as the other. How would the little explorers deal with this?

As you might expect, the ants quickly determined which branch was best (this is the same species, after all, that demonstrates such a knack for locating maple syrup spilled on your kitchen floor). In most trials of the experiment, after an initial period of wandering, all of the ants chose the shorter branch.

The pheromone trail was the key. As more and more ants picked the shorter branch, it accumulated more and more of their pheromone, increasing the likelihood that other ants would choose it. Here's how it works: Let's say two ants set out across the bridge at the same time. The first ant takes the shorter branch, and the second the longer one. By the time the first ant reaches the food, the second is only halfway across the bridge. By the time the first ant returns all the way to the colony, the second ant has just arrived at the food. To a third ant standing at the split in the bridge at this point, the pheromone trail left by the first ant would be twice as strong as that left by the second (since the first ant went out and returned), making it more likely to take the shorter branch. The more this happens, the stronger the pheromone trail grows, and the more ants follow it.

Ant colonies, in other words, have evolved an ingenious way to determine the shortest path between two points. Not that any of the ants are doing so on their own. None of them attempts to compare the length of the two branches independently. Instead, the colony builds the best solution as a group, one individual after another, using pheromones to "amplify" early successes in an impressive display of self-organization.

Taking this idea one step further, Deneubourg and his colleagues proposed a relatively simple mathematical model to describe this behavior. If you know how many ants have taken the shorter branch at any particular time, Deneubourg said, you can reliably calculate the probability of the next ant choosing that branch. To demonstrate this, he plugged his team's equations into a computer simulation of the double-bridge experiment and ran it for a thousand ants. The results mirrored those of real ants. When the branches were the same length, the odds of an ant picking either one were fifty-fifty. But when one branch was twice as long as the other, the odds of picking the shorter one shot up dramatically.

The key to the colony's system, in short, lay in the simple rules that each ant applied to local information. If you changed these rules, you would change the behavior pattern of the whole colony.

The implications of this discovery were not lost on Dorigo: If real ant colonies could find the shortest path between two points, then why couldn't researchers do the same thing with "virtual ants"? Dorigo knew how to design software "agents" that could follow simple rules just like real ants do. Why couldn't these software agents find the shortest path, too? Only, what if the path wasn't the distance between an ant nest and a pile of food? What if it was the shortest route for a message across the Internet between two computers? Or the shortest distance for a package going from a factory warehouse in California to a customer in Florida? Or the shortest path between multiple steps in an industrial process? Then there was the concept of the "shortest path" itself. What if you redefined "shortest" as "most efficient" or "least costly"? Wouldn't that be a handy tool?

"When I went back to Milan to discuss these ideas with Professor Alberto Colorni, who was supervising my work, he asked me to write a simple program as a proof of principle, to show it wasn't just a crazy idea," Dorigo says. At the time, Dorigo was working on a class of mathematical puzzles known as combinatorial optimization problems, which are relatively easy to describe but deceptively difficult to solve. One of the best-known examples is the traveling salesman problem, which involves the following scenario: A salesman needs to visit customers in a number of cities. What's the shortest path he can take to visit each city once before returning back home?

When the problem involves just a few cities—let's say, Moscow, Hong Kong, and Paris—you can figure out the answer on the back of an envelope. Leaving from the airport near his home in Cleveland, the salesman has three options for his first stop: Moscow, Hong Kong, or Paris. Let's say he chooses Hong Kong. From there he has two choices: Moscow or Paris. Let's say he flies to Paris. That leaves only Moscow before he can fly home. If you made a list of all the other possible sequences (such as Paris to Moscow to Hong Kong, or Moscow to Hong Kong to Paris, and so on), you would have a total of six to consider. Compare the mileage for each sequence and you have your answer.

But here's the tricky part. If you add a fourth city to the salesman's journey, you make the problem significantly more difficult. Now you have four times as many possible routes to consider—twenty-four instead of six. Add a fifth city and you get 120 possible routes. Jump ahead to ten cities and you're talking about more than 3.6 million possible routes. The number of solutions, in other words, increases exponentially with each new city. When you get to thirty cities, there aren't enough years in a lifetime to list all the possible routes.

Dorigo thought it might be interesting instead to let virtual ants give it a try. Rather than trying to identify every possible outcome of the salesman's decision making, the ant system used trial-and-error shortcuts to find a handful of good ones. Instead of being straightforward and linear, it was decentralized and distributed. Instead of calling for complicated calculations, it relied on simple rules of thumb. Instead of getting swamped by the exponential nature of the math, it took advantage of that same snowballing effect to rapidly turn small differences into big advantages. It was different, in other words, because it harnessed self-organization in a smart way.

So Dorigo and Vittorio Maniezzo, a fellow graduate student, created a set of virtual ants capable of cooperating with one another to find the shortest route for the traveling salesman. Their secret weapon: "virtual pheromones" that the ants would leave along the way. Imagine a map with fifteen cities that a salesman needed to visit. At the beginning of the first cycle, ants were placed randomly on all of the cities. Then each ant used a formula based on probability to decide which town to visit next. This formula considered two factors: which city was closest, and which city had the strongest pheromone trail leading to it. At the start, there were no pheromone trails, so the closest cities tended to be selected. As soon as each ant completed a tour of all fifteen cities, it retraced its path, depositing virtual pheromone on its route. The shortest routes discovered by the ants tended to receive the most pheromone, while the longest ones were allowed to "evaporate" more rapidly. This enabled the ants, as a group, to remember the best routes. So when the second cycle was run, and the ants worked their way from city to city again, each ant built upon the successes of the first cycle by favoring the strongest pheromone trails. After repeating this time and again, the ants kept reducing their travel times, until the pheromone trails on the shortest segments were so strong that none of them could resist choosing them.

The results were quite encouraging.

"We discovered that the ants could find nearly optimal solutions for as many as thirty, fifty, or even a hundred cities," Dorigo says.

Not that the ants didn't sometimes make mistakes. If a particular ant, while hopping from city to city, happened to get caught in a loop along the way—like a twig in a river eddy—other ants occasionally followed, resulting in a nonsolution to the problem. To prevent this, Dorigo and Maniezzo instructed the ants to forget such loops when depositing pheromone on completed trails. Other problems required similar fixes. But none of the ants' bad habits were so serious that they couldn't be tweaked in one way or another, or made more effective by pairing them with more specialized algorithms.

"The important thing was that the ant colony optimization worked because of the implicit cooperation among many agents," Dorigo says. "On their own, each artificial ant built a solution which was usually not very good. But together, by exchanging information—not talking to each other, but simply by exchanging information through virtual pheromones—the ants ended up finding some very good solutions." In this way, cooperation became cumulative. Instead of representing simply the sum of the ants' individual efforts, the search got smarter as it went forward, powered by the mechanisms of self-organization—*decentralized control*, *distributed problem-solving*, and *multiple interaction between agents*.

It wasn't long before other computer scientists were adapting the ant-colony approach to solve a variety of difficult problems. A few researchers even experimented with real-world situations. At Hewlett-Packard's lab in Bristol, England, for example, scientists created software to speed up telephone calls. Using a simulation of British Telecom's network, they dispatched antlike agents into the system to leave pheromone-like signals at routing stations, which function as intersections for traveling messages. If a station accumulated too much digital pheromone, it meant that traffic there was too congested, and messages were routed around it. Since the pheromone evaporated over time, the system was also able to adapt to changing traffic patterns as soon as congested stations opened up again.

What did an ant-based algorithm offer that other techniques didn't? The answer goes back to the foragers from Colony 550. If conditions in the desert changed while the ants were out searching for seeds—if something unpredictable disrupted the normal flow of events, such as a hungry lizard slurping down ants—then the colony as a whole reacted quickly: foragers raced back to the nest empty-handed; other ants didn't go out. They didn't wait for news about the disruption to travel up a chain of command to a manager, who evaluated the situation before issuing orders that traveled back down the chain to workers, as might happen in a human organization. The colony's decision making was decentralized, distributed among hundreds of foragers, who responded instantly to local information. In the same way, virtual ants racing through the telephone network responded instantly to congested traffic. In both cases, an ant-based algorithm offered a flexible response to an unpredictable environment, and it did so using the principles of self-organization.

An obvious application of this approach would be to develop an algorithm—or set of algorithms—that would enable a business enterprise to respond to changes in its environment as quickly and effectively as ant colonies do. That's exactly what a company in Texas set out to accomplish.

The Yellow Brick Road

Charles Harper looked out his office window at the flat landscape south of Houston. As director of national supply and pipeline operations at American Air Liquide, a subsidiary of a $12 billion industrial group based in Paris, he supervised a team monitoring a hundred or so plants producing medical and industrial gases. This was a daunting task on the best of days. The company's operations were so complex that no two situations ever looked the same.

Air Liquide sold different types of gas to a wide range of customers. Hospitals bought oxygen, as did paper mills and plastics manufacturers. Ice cream makers used liquid nitrogen to freeze their goods. So did berry packers and crawfish shippers. Soft drink companies purchased carbon dioxide to add fizz to their beverages. Oil refineries took several gases, as did steel mills. All told, Air Liquide delivered gas products to more than fifteen thousand customers across the United States, using a fleet of seven hundred trucks, three hundred rail cars, and a 2,200-mile network of pipelines.

All these moving parts, however, were just the beginning of the business problem. The real complexity came from the variables the company had to cope with. The cost of energy, for example, fluctuated constantly. In Texas, where the power industry was deregulated in 2002, the price of electricity changed every fifteen minutes. "For an industrial customer, a megawatt might cost $18 at three a.m., then shoot up to $103 the following afternoon," Harper says. Since energy was one of Air Liquide's biggest expenses, accounting for up to 70 percent of the cost of production, these ups and downs had a huge impact on the bottom line.

Other factors affected production costs. Each of the plants producing gaseous or liquid gases had a different efficiency level, different cost profile, and different capacity. Many, for example, could produce either liquid oxygen or liquid nitrogen in varying combinations. For customers who needed delivery by truck, a plant could pump liquid gases into cryogenic trailers. For those on pipelines, it could vaporize the gases and send them that way.

Customer demand was yet another variable. Although some customers, usually the largest ones, took the same amount of gas every week, many others were unpredictable. A small company might order gas only when it got a big contract, then order none for months. About 20 to 30 percent of Air Liquide's customers made a habit of calling in special requests. "If a big medical center calls us up and says, hey, we need a delivery of oxygen right away, we're going to make sure they don't run out," Harper says. But such requests put a strain on scheduling.

Combine all these factors—fluctuating energy prices, changing production costs, varying delivery modes, and uncertain customer demand—and you've got a difficult situation to manage. Sooner or later, something unpredictable, like a mechanical problem at a plant, is going to put you in a bind, and you won't have enough gas to serve customers in that region. "We were always having incidents like that," says Clarke Hayes, Air Liquide's real-time operations manager. "It finally got to the point where we said, you know what, we need a tool that helps us organize better."

The company already had special-purpose programs to optimize particular aspects of their operations, but it didn't have a way to pull it all together. In late 1999, a team from Bios Group, a consulting firm from Santa Fe, New Mexico, founded by complexity scientists, came to Air Liquide with an unorthodox proposal. Why not build a computer model based on the self-organizing principles of an ant colony? This model, they suggested, would take into account all the variables that were making planning so difficult as a way to help managers find solutions to day-to-day challenges. As a start, they suggested tackling the company's truck-routing problem—the question of which truck should pick up gas from which plant and deliver it to which customer to be most profitable for the company. If ants had evolved a clever way to move things from one place to another, they said, why not apply that knowledge to Air Liquide's trucks?

"The scientists were wonderful to talk to," Harper says. "But the issue for us was, can they understand the industrial gas business? So we took a small piece of our geography and asked them to digitize that. To show us they understood the complexity of the trucks, the drivers, the depot costs, the miles per gallon, all the anomalies. What if a customer's tank was on a hill? If you pull up in the wrong direction, or if your truck's not full, the liquid won't get in the pump and you can't fill the tank. So you have to make that customer the first stop on your route. There are hundreds of those kinds of things, and they drive you crazy. But they all needed to be in the model."

Alberto Donati was one of the scientists at Bios Group assigned to work on the Air Liquide pilot project. Because he had previous experience with ant-based algorithms, he was asked to work on the distribution side of the decision-support system. The approach he took was inspired by the one Marco Dorigo and Eric Bonabeau, another computer scientist, had developed for the traveling salesman problem and similar difficult problems.

"The ant algorithm was a very good choice in this case, because it creates a step-by-step procedure to find the best routing solution," Donati says. At every step, even the most complex situation could be taken into account. Each ant had a sort of "to do" list that it kept working on until the list was complete, he explained. Let's say the list was of Air Liquide customers that need deliveries today. "Imagine the ant starts at the depot," Donati says. "First she has to choose a truck. So she looks at the available list of trucks, and then she picks a driver. So what does she do next? Maybe she goes to the facility to fill up the truck. Now she considers all the possible customers that need that kind of gas. She calculates the time it would take to reach each customer's site. Perhaps there are some customers with restricted time windows for deliveries, or others with high priority for deliveries. Then the ant looks at each customer using what we call a *greedy function*." The term *greedy*, in this case, refers to a decision-making rule that delivers the best results in a short time frame. "Choose the nearest customer," for example, is a typical greedy function. "She also takes into account the pheromone trail," Donati says. "Other ants may have chosen the same path and left some pheromone. So she multiplies the greedy factors by the pheromone factor to determine which customer to choose next." (This decision is modified by a small degree of randomness to occasionally allow choices that would be hard to predict.) "When she gets to the customer, she unloads the needed amount of the truck's liquid, keeping track of how much time it takes to do that and how much is left in the truck. Then she goes back to the list of customers she hasn't visited yet." And so on, and so on, until all the customers have been visited and assigned to a route.

At this point, the ant computes the quality of the solution and lays down a pheromone trail according to its quality. This process is repeated, ant after ant, thousands of times. "The nice part is that, when you're near the finish, you will see that the ants have left a clear distribution of pheromones around your system,"

Donati says. Each new solution is compared to the best previous one. If it's better, it becomes the best one. It's all a matter of balance between exploration and exploitation, he says.

The pilot project was a big hit at Air Liquide, proving to managers that an ant-based model was flexible enough to handle the complexities of their routing problem. But what Air Liquide really wanted was to optimize production, since the cost of producing gas was ten times that of delivering it. So they enlisted Bios Group, which by then had merged with a company called NuTech Solutions, to develop a tool to optimize production. That tool, completed in 2004, is the one Air Liquide uses to guide its business today.

Technicians in the control center run this optimizer every night. They begin at eight p.m. by entering new data about plant schedules, truck availabilities, and customer needs into the model. A telemetry-based system called SCADA (Supervisory Control and Data Acquisition) feeds real-time information about the efficiency of each plant, gas levels in storage tanks, and the cost of power, among other factors. A neural network forecast engine provides estimates of which customers must get deliveries right away, based on telemetry readings and previous customer-usage patterns. Weather forecasts by the hour are entered, as are estimated power costs for the next week. Finally, any miscellaneous information is added that might affect schedules, such as which plants need maintenance in the near future.

The optimizer is then asked to consider every permutation—millions of possible decisions and outcomes—to come up with a plan for the next seven days. To do so, it combines the ant-based algorithm with other problem-solving techniques, weighing which plants should produce how much of which gas. To speed up run times, technicians divide the country into three regions: west of the Rockies, Gulf Coast, and the eastern states. Then they run the model three times for each region. By the time the day crew arrives at work at six a.m., the optimizer has solutions for each region.

People still make all the decisions. But now, at least, they know where they need to go. "One of the scientists called this approach the Yellow Brick Road," Harper says. "Basically, instead of worrying about the absolute answer, we let the optimizer point us toward the right answer, and by the time we take a few more steps, we rerun the solutions and get the next direction. So we don't worry about the end point. That's Oz. We just follow the Yellow Brick Road one step at a time."

This ant-inspired system has helped Air Liquide reduce its costs dramatically, primarily by making the right gases at the right plants. Exactly how much, company officials are reluctant to say, but one published estimate put the figure at $20 million a year.

"It's huge," Harper says. "It's actually huge."

Lessons from Checkers

During the 1950s, an electrical engineer at IBM named Arthur Samuel set out to teach a machine to play checkers. The machine was a prototype of the company's first electronic digital computer called the Defense Calculator, and it was so big it filled a room. By today's standards, it was a primitive device, but it could execute a hundred thousand instructions a second, and that was all that Samuel needed.

He chose checkers because the game is simple enough for a child to learn, yet complicated enough to challenge an experienced player. What makes checkers fun, after all, is that no two games are likely to be exactly the same. Starting with twelve pieces on each side and thirty-two squares on the game board to choose from (checkers is played only on the dark squares), the number of possible board configurations from start to finish is practically endless. You can play over and over and never repeat the same sequence of moves. This gives checkers what complexity experts call *perpetual novelty*.

For Samuel's computer, that was a problem. If every move theoretically could lead to billions of possible configurations of the game board, how could it choose the best one to make? Compiling a comprehensive list of results for each move would simply take too long—just as it would for Marco Dorigo in the traveling salesman problem. So Samuel gave the machine a few basic features to look for. One was called *pieces ahead*, meaning the computer should count how many pieces it had left on the board and compare that with its opponent's. Was it two pieces ahead? Three pieces ahead? If a particular move resulted in more pieces ahead, it was likely to be favored. Other features specified favorable regions of the board. Penetrating the opponent's side was considered advantageous, for example. So was dominating the middle. And so on.

Samuel also taught the computer to learn from its mistakes. If a move based on certain features failed to produce a favorable outcome, then the computer gave less weight to those features the next time around. In addition, he showed the computer how to recognize "stage-setting" moves—those that didn't help out in an obvious way right now, such as a move that sacrificed a piece, but set up a later move with a bigger payoff, such as a triple jump. The machine did this after the fact by increasing the weight of features that favored the stage-setting move. Finally, he told the computer to assume that its opponent knew everything that it knew so the opponent would inflict the greatest damage possible whenever it could. That forced the machine to factor in potentially negative consequences of moves as well as positive ones. If it got surprised by an opponent anyway, it adjusted the weights to avoid that mistake next time.

Samuel's project was so successful that the computer was soon beating him on a regular basis. By the end of the 1960s, it was defeating checkers champions.

"All in all, his was a remarkable achievement," writes John Holland, another pioneer of artificial intelligence, in his book *Emergence: From Chaos to Order*. "We are nowhere near exploiting these lessons that Samuel put before us almost a half century ago."

To Holland, who shared a lab with Samuel at IBM, the true genius of the checkers program was the way it modified the weights of a handful of features to cope with the game's daunting complexity. Because it was impractical at the time to "solve" the game of checkers mathematically by calculating the perfect sequence of moves, as you might do with a simpler game, such as tic-tac-toe, Samuel just wanted his computer to play the game *better* each time. "The *emergence* of good play is *the* objective of Samuel's study," Holland wrote.

What Holland meant by *emergence* was something quite specific. He was referring to the process by which the computer formed a master strategy as a result of individual moves, or, as he put it more generally, the phenomenon of *much coming from little*. Although everything the program did was "fully reducible to the rules (instructions) that define it," he says, the behaviors generated by the game were "not easily anticipated from an inspection of those rules."

We saw the same thing, of course, in Colony 550. Even though individual ants were following simple rules about foraging, their pattern of behavior as a group added up to a surprisingly flexible strategy for the colony as a whole. One colony might tend to be more aggressive in its style of foraging, sending out lots of foragers, while another might be more conservative, keeping them safe inside. Each colony didn't impose its strategy on the foragers; the strategy *emerged* from their interactions with one another.

The same could be said about many complex systems, from beehives and flocks of birds to stock markets and the Internet. Whenever you have a multitude of individuals interacting with one another, there often comes a moment when disorder gives way to order and something new *emerges*: a pattern, a decision, a structure, or a change in direction. This whole chapter, in fact, has been about the kinds of strategies that emerge from self-organized behavior. And what these strategies all have in common is that they represent a way to cope with the unpredictable.

Consider life in an ant colony, where survival means competing not only against other colonies but also against an ever-changing environment. Will there be enough food today? Where will it be found? How will the weather affect the nest? The colony meets such challenges through self-organized behavior, and what emerges is a pattern of activity that allocates the colony's resources to meet its immediate needs.

Air Liquide, for its part, had its own list of unknowns. Which customers would need deliveries today? What types of gas would they need? Which production facilities could make those gases at the least cost? What would the price of electricity be at those facilities? How could the company deliver those gases most economically? By emulating an ant colony's distributed problem-solving approach, the company's optimizer tool provided a day-to-day plan to cope with an endless string of variables.

Like many businesses today, Air Liquide was looking for a way to cope with the perpetual novelty of its environment. The company didn't expect a guarantee, that it would win every competition it got into, just an opportunity to stay in the game until it could adapt to the latest changes. What it needed, in other words, was a strategy to gain a degree of control over the uncontrollable—which was what Samuel's checkers player also seemed to promise.

That was quite different, in an important way, from what Deborah Gordon's ant colonies were trying to do. Instead of attempting to outsmart the desert environment, the ants, in a sense, were matching its complexity with their own. If Colony 550 were to play a game of checkers, each piece on the board would move by itself, acting on local information, with nobody waiting for instructions. The game would be a swirl of motion as pieces moved forward, jumped over one another, became kings, or got taken as prisoners in patterns of interactions that might be difficult to perceive at first glance. But if checkers were as important to ants as foraging, the colony, without doubt, would be a flexible and resilient competitor.

This tension between minimizing uncertainty, on the one hand, and experimenting to keep up with change, on the other, is something we'll see time and again throughout this book. And what's surprising about the behavior evolved by bees, birds, and fish, among other species, is the adroit way that groups of such animals manage to have it both ways—to manage complexity and to partake of it at the same time.

"If I was in charge of designing the software for a company like Air Liquide, I'd probably be stressed about doing a really great job," Gordon says. "But the ants aren't doing that." Their system's too loose and undisciplined. Information coming in is too spotty, and their responses are too unpredictable. "The amazing thing to me is how, every way you look at it, the ants' system is so messy, and yet somehow it works," she says.

Maybe there's a deeper lesson here, Gordon suggested. "Instead of trying to keep fine-tuning a system so it will work better and better, maybe what we really ought to be looking for is a rigorous way of saying, okay, that's good enough." Maybe a smart way to face the unpredictable, whether you're running a business or playing a game of checkers, is to look for that balance between strategic goals and random experimentation. Ant colonies, after all, manage to thrive at the edge between efficiency and utter chaos, she says. "The question is, how do they find that edge? Because if we could find that edge too, we could save ourselves a lot of trouble."

2

Honeybees

Making Smart Decisions

Appledore Island is a tough place for honeybees. Anchored in the Atlantic off the coast of southern Maine, the rocky, wind-blown island is barely a half-mile long, with hardly any trees, which the bees need for nest building. In fact, you might describe the island as a kind of bee Alcatraz, which makes it an ideal place to observe their behavior under controlled conditions.

A few summers ago, biologists Thomas Seeley of Cornell University and Kirk Visscher of the University of California at Riverside ferried a half-dozen colonies of honeybees to Appledore, which is home to the Shoals Marine Laboratory run by Cornell. For nearly a decade, Seeley and Visscher have been studying a fascinating example of what they call "animal democracy." How do several thousand honeybees, they want to know, put aside their differences to reach a decision as a group?

The focus of their research has been honeybee "house hunting." In late spring or early summer, as a large hive outgrows its nest, the group normally divides. The queen and roughly half of the bees fly off in a swarm to create a new colony, leaving behind a daughter queen in the old nest. There may be fifteen thousand bees in the swarm, which typically clusters on a tree branch, while several hundred scout bees search the neighborhood for new real estate. Although the queen's presence is important to bees in the swarm, she plays no role in picking a new nest site. That task is delegated to the scouts, who do their jobs without direction from a leader.

When a scout buzzes off into the countryside, she's looking for just the right dwelling place (I say *she*, because worker honeybees are all females). It must be well off the ground, with a small entrance hole facing south and enough room inside to allow the colony to grow. If she finds such a spot—a hollow in a tree would be perfect—she returns to the swarm and reports her discovery by doing a waggle dance. This dance, which resembles the one forager bees do when they locate a new patch of flowers, contains a code telling others how to find the site. Some of the scouts that see her dance will then go examine the site for themselves, and, if they agree with her assessment, they'll return to the swarm and dance in support of the site, too.

This is no trivial question for the bees. As long as the swarm is clinging to the branch, it remains exposed to weather, predators, and other hazards. But once the swarm selects a new home, it won't move again until next spring. So it has to get it right the first time. If the group selects poorly, the entire colony could perish.

One by one, scouts that have been exploring the neighborhood return to the cluster with news about different locations. Soon there's a steady stream of bees flying between the cluster and a dozen or more potential nest sites, as more and more scouts get involved in the selection process. Eventually, after enough scouts have inspected enough sites, it becomes clear that traffic at one site is much greater than that at any other, and a decision is reached. The bees in the main cluster warm up their wings and fly off together to the chosen site—which almost always turns out to be the best one.

Facing a life-or-death situation, in other words, a honeybee swarm engages in a complex decision-making process involving multiple, simultaneous interactions between hundreds of individuals with no leadership at all—exactly the kind of chaotic, unpredictable enterprise that, if attempted by people under stress, would almost certainly lead to disaster. Yet the bees almost always make the right choice.

How do they do it?

The Five-Box Test

One spring day in 1949, a young zoologist named Martin Lindauer was observing a swarm of bees near the Zoological Institute in Munich, Germany, when he noticed something odd. Some of the bees, he realized, were doing waggle dances. Ordinarily that meant they were foragers that had found a nice patch of flowers nearby, and they were telling other bees where to go find it. But these dancers weren't carrying any pollen or nectar, so Lindauer didn't think they were foragers. What were they up to?

Lindauer's mentor at the University of Munich, the renowned zoologist Karl von Frisch, had recently figured out that the waggle dance—or "tail-wagging dance" as he called it—was in fact a sophisticated form of communication (he won a Nobel Prize for this research in 1973). When a foraging bee danced, von Frisch had discovered, she wasn't just advertising a source of food, she was also providing precise directions to locate it. To perform such a dance, a bee would run forward a short distance on the hive's comb while vibrating her abdomen in a "waggle." Then she'd return to her starting point in a figure-eight and repeat this over and over, as if reenacting her flight to the flower patch. The length of her dance indicated how far away the food was, and the angle of her dance (relative to vertical) corresponded with the direction of the food (relative to the sun). If a bee danced in a direction thirty degrees to the right of vertical, for example—picture the number 1 on a clock—the flower patch could be found by flying in a direction thirty degrees to the right of the sun. It was an ingenious system, but it had never been linked to house hunting before.

By carefully studying several swarms—sometimes running beneath them as they flew across the Bavarian countryside—Lindauer determined that the bees dancing on the swarm cluster were scouts that had been out searching for a nest site. Some were still powdered with red-brick dust from having explored a hole in a building, or blackened with soot from having checked out a chimney. Just as foraging bees used the waggle dance to share news about food sources, so the scouts were using it to report on potential real estate. At first, many of the scouts danced in different directions, apparently announcing various options. But after some hours, fewer and fewer sites were mentioned until, finally, all the dancers were pointing in the same direction. Soon after that, the swarm lifted off from its bivouac and flew to its new home, which Lindauer was able to locate by reading the code of the dances.

The bees had reached a consensus, he theorized, because the liveliest scouts had persuaded the rest to go along with their choice. They did this by getting rivals to visit their preferred site, where, confronted with the superior qualities of the site, the former competitors simply changed their minds. One by one they were won over, he speculated, and the disagreement went away.

In this respect, at least, Lindauer got it wrong. It wasn't quite that simple. Researchers have since established that only a small percentage of scouts ever visit more than one site. The group's decision does not rely on individual scouts changing their minds, but rather on a process that combines the judgments of hundreds of scouts—one that would remain a mystery for fifty years.

That's where Tom Seeley and Kirk Visscher came in. Beginning in the late 1990s, they picked up where Lindauer left off, this time using video cameras to record every aspect of the swarm's behavior. They also brought some new ideas about honeybee deliberation. Given the large number of individuals that take part in house hunting, they doubted that bees' decision making was based on consensus. It just seemed too complicated, like trying to get a large group of friends to agree on which movie to watch. More likely, they figured, the process relied on some form of competition. Instead of trying to work through their differences with one another, scouts dancing on the swarm cluster appeared to be actively lobbying for different sites. It wasn't a meeting of minds at all, but a race to build up supporters—with the winners taking all.

In that sense, the bees' system was more like a stock market, in which the value of a security rises or falls according to the collective judgment of the group. Scouts watching another scout dance, like brokers, might be persuaded to do their own research on the site being advertised, and if they liked what they saw, they could buy into the site by dancing for it themselves. If they didn't like it, they didn't have to. The more bees that joined in, the greater the likelihood the site would be selected.

But how did the process work, exactly? What were the mechanisms that enabled the bees to choose so accurately?

To find out, Seeley and Visscher conducted a series of experiments. After preparing a swarm for house hunting, they placed five plywood nest boxes an equal distance from the bees on Appledore Island—four representing mediocre choices for a new nest and one that was excellent. What made the fifth box better than the rest was that it offered the bees an ideal amount of living space—about forty quarts, compared to fifteen quarts for the others, which was not enough to store honey, raise brood, and meet the other needs of an expanding colony. To track the bees during the decision-making process, Seeley and Visscher labeled all four thousand individuals in each swarm with tiny numbered disks on their thoraxes and dabs of paint on their abdomens, a tedious process that involved chilling batches of twenty bees at a time to render them docile enough to be handled. But it was worth it in the end, because, when they looked at video tapes of the swarms later, they could tell which bees had visited which nest boxes and which ones had danced for which boxes at the main cluster. The shape of the decision-making process emerged.

The key, it turned out, was the brilliant way the bees exploited their *diversity of knowledge*—the second major principle of a smart swarm. Just as Deborah Gordon's ant colonies used *self-organization* to adjust to changes in the environment, so the honeybees used diversity of knowledge to make good decisions. By diversity of knowledge, in this case, I mean a broad sampling of the swarm's options. The more choices, the better. By sending out hundreds of scouts at the same time, each swarm collected a wealth of information about the neighborhood and the nest boxes, and it did so in a distributed and decentralized way. None of the bees tried to visit all five of the boxes to rate which one was the best. Nor did they submit their findings to some executive committee for a final decision, as workers in a corporation might do. Instead, these hundreds of scouts each provided unique information about the various sites to the group as a whole in what Seeley and Visscher described as a "friendly competition of ideas."

Equally important, every scout evaluated nest sites for herself.

If a scout was impressed by another scout's dance, she might fly to the box being advertised and conduct her own inspection, which could last as long as an hour. But she would never blindly follow another scout's opinion by dancing for a site she hadn't visited. That would open the door to untested information being spread like a rumor. Or, to use the stock-broker analogy, a bee wouldn't invest in a company just because its stock was on the rise. She'd check out its fundamentals first.

Meanwhile, as the scout bees continued their search, the swarm was busily ranking each option. This was determined by the number of bees visiting each site. The more visitors, the more "votes" for the site. Though the best nest box wasn't discovered first by the scouts, it quickly attracted the attention of numerous bees. Scouts returning from the excellent box had no trouble convincing others to check it out, largely because they danced for it so vigorously—performing as many as a hundred dance circuits each, compared to only a dozen or so danced by bees for lesser sites. A dance of that length could take five minutes, compared with thirty seconds for a shorter dance, so it was much more likely to be noticed by scouts walking around on the surface of the cluster. And once the number of bees advertising the best box increased, support for it shot up, as interest in the mediocre sites faded away.

"This careful tuning of dance strength by the scouts created a powerful positive feedback," Seeley explained, "which caused support for the best site to snowball exponentially." This was a crucial mechanism, because it meant that even small differences in the quality of nests were exaggerated—their "signals" were amplified—making it much more likely that support for the best site would surge ahead.

As more and more bees gathered at the first-rate box, fewer and fewer lingered at the others. That was because scouts returning from boxes for the second or third time were dancing fewer circuits for them each time, whether they'd visited the excellent box or the mediocre ones. Scouts that had visited poor sites quit dancing first. Seeley and Visscher described this mechanism as the dance "decay rate." It meant that support for less attractive boxes would dwindle automatically—even as the number of bees collecting at the superior box kept growing—in a decision-making process that lasted from two to five hours during the test. In technical terms, this represented a balancing, or negative feedback, preventing the swarm from choosing too fast and making a mistake. These were the factors steering the bees' problem-solving machine—exponential recruitment on the accelerator, dance decay rate on the brakes.

Meanwhile, something critical was happening at the nest boxes. As soon as the number of bees visible near the entrance to the best box reached fifteen or so, Seeley and Visscher noticed a new behavior among the scouts. Those returning from the box started plowing through bees in the main cluster, producing a special signal called "worker piping."

"It sounds like *nnneeeep, nnneeeep!* Like a race car revving up its engine," Seeley says. "It's a signal that a decision has been reached and it's time for the rest of the swarm to warm up their wing muscles and prepare to fly." Scouts from the excellent box, in other words, were announcing that a quorum had been reached. Enough bees had "voted" for the most attractive box by gathering there at the same time. A new home had been chosen.

The number fifteen, it turns out, was the threshold level for the quorum. Although this number might seem arbitrary at first glance, it turns out to be anything but that. Like the dance decay rate, the threshold level represents a finely tuned mechanism of emergence. To gather that many bees at the entranceway simultaneously, it takes as many as 150 scouts traveling back and forth between the box and the main swarm cluster, which means that a majority of the bees taking part in the selection process have committed themselves to the site.

Once the quorum was reached, the final step was for scouts to lead the rest of the group to the chosen site. Most of the swarm, some 95 to 97 percent, had been resting during the whole decision-making process, conserving their energy for the work ahead. Now, as the scouts scrambled through the crowd, they stopped from time to time to press their thoraxes against other bees to vibrate their wing muscles, as if to say, warm up, warm up, get ready to fly. A final signal, called the buzz run, in which the scouts bulldoze through sleepy workers and buzz their wings dramatically, triggered the takeoff. At that point, the whole swarm flew away to its new home—which, to nobody's surprise, turned out to be the best nest box.

The swarm chose successfully, in short, because it made the most of its diversity of knowledge. By tapping into the unique information collected by hundreds of scouts, it maximized its chances of finding the best solution. By setting the threshold level high enough to produce a good decision, it minimized its chances of making a big mistake. And it did both in a timely manner under great pressure to be accurate.

The swarm worked so efficiently, in fact, you might be tempted to imagine it as a complicated Swiss watch, with hundreds of tiny parts, each one smoothly performing its function. Yet the reality is much more interesting. To watch a swarm in the midst of deliberation is to witness a chaotic scene not unlike the floor of a commodities market, with dozens of brokers shouting out orders at the same time. Bees coming and going. Scouts dancing this way or that. Uncommitted bees milling around. The way they make decisions looks very messy, which is also very beelike. Natural selection has fashioned a system that is not only tailor-made for their extraordinary talents for cooperation and communication but also forgiving of their tendency to be unpredictable. It is from this controlled messiness that the wisdom of the hive emerges.

Seek a diversity of knowledge. Encourage a friendly competition of ideas. Use an effective mechanism to narrow your choices. These are the lessons of the swarm's success. They also happen be the same rules that enable certain groups of people to make smart decisions together—from antiterrorism teams to engineers in aircraft factories—through a surprising phenomenon that has come to be known as the "wisdom of crowds."

The Wisdom of Crowds

In early 2005, Jeff Severts, a vice president at Best Buy, decided to try something different. Severts had recently attended a talk by James Surowiecki, whose bestseller *The Wisdom of Crowds* claims that, under the right circumstances, groups of nonexperts can be remarkably insightful. In some cases, Surowiecki argues, they can be even more intelligent than the most intelligent people in their ranks. Severts wondered if he might be able to tap into such braininess at Best Buy. As an experiment, in late January 2005 he sent e-mails to several hundred employees throughout the company, asking them to predict sales of gift cards in February. He got 192 replies. In early March, he compared the average of these estimates to actual sales for the month. The collective estimate turned out to be 99.5 percent accurate—almost 5 percent better than the figure produced by the team responsible for sales forecasts.

"I was surprised at how eerily accurate the crowd's estimates were," Severts says.

In his book about smart crowds, Surowiecki cites similar examples of otherwise ordinary people making extraordinary decisions. Take the quiz show *Who Wants to Be a Millionaire?* Contestants stumped by a question are given the option of telephoning an expert friend for advice or of polling the studio audience, whose votes are averaged by a computer. "Everything we know about intelligence suggests that the smart individual would offer the most help," Surowiecki writes. "And in fact the 'experts' did okay, offering the right answer—under pressure—almost 65 percent of the time. But they paled in comparison to the audiences. Those random crowds of people with nothing better to do on a weekday afternoon than sit in a TV studio picked the right answer 91 percent of the time."

Although Surowiecki readily admits that such stories by themselves don't amount to scientific proof, they do raise a good question: If hundreds of bees can make reliable decisions together, why should it be so surprising that groups of people can too? "Most of us, whether as voters or investors or consumers or managers, believe that valuable knowledge is concentrated in a very few hands (or, rather, in a very few heads). We assume that the key to solving problems or making good decisions is finding that one right person who will have the answer," Surowiecki writes. But often that's a big mistake. "We should stop hunting and ask the crowd (which, of course, includes the geniuses as well as everyone else) instead. Chances are, it knows."

Severts was so impressed by his first few efforts to harness collective wisdom at Best Buy that he and his team began experimenting with something called *prediction markets*, which represent a more sophisticated way of gathering forecasts about company performance from employees. In a prediction market, an employee uses play money to bid on the outcome of a question, such as "Will our first store in China open on time?" A correct bid pays $100, an incorrect bid pays nothing. If the current price of a share in the market for a bid that yes, the store will open on time, is $80, for example, that means the entire group believes there's an 80 percent chance that that will happen. If an employee is more optimistic, believing there's a 95 percent chance, he might take the bet, seeing an opportunity to earn $15 per share. In the case of the new store, which had been scheduled to open in Shanghai in December 2006, the prediction market took a dive, falling from $80 a share to $50 eight weeks before the opening date—even though official company forecasts at the time were still positive. In the end, the store opened a month late.

"That first drop was an early warning signal," Severts says. "Some piece of new information came into the market that caused the traders to radically change their expectations." What that new information might have been about, Severts never found out. But to him it didn't really matter. The prediction market had proved its ability to overcome the many barriers to effective communication in a large company. If anyone was listening, the alarm bells were ringing loud and clear.

As this story suggests, there may be several good reasons for companies to pay attention to prediction markets, which are good at pulling together information that may be widely scattered throughout a corporation. For one thing, they're likely to provide unbiased outlooks. Since bids are placed anonymously, markets may reflect the true opinions of employees, rather than what their bosses want them to say. For another thing, they tend to be relatively accurate, since the incentives for bidders to be correct—from T-shirts to cash prizes—encourage them to get it right, using whatever unique resources they might have.

Above and beyond these factors is the powerful way prediction markets leverage the simple mathematics of *diversity of knowledge*, which, when applied with a little care, can turn a crowd of otherwise unremarkable individuals into a comparative genius. "If you ask a large enough group of diverse, independent people to make a prediction or estimate a probability, and then average those estimates, the errors each of them makes in coming up with an answer will cancel themselves out," Surowiecki explains. "Each person's guess, you might say, has two components: information and error. Subtract the error, and you're left with the information."

The house-hunting bees demonstrate this math very clearly. When several scouts return to the swarm from checking out the same perfect tree hollow, for example, they frequently give it different scores—like opinionated judges at an Olympic ice-skating competition. One bee might show great enthusiasm for such a high-quality site, dancing fifty waggle runs for it. Another might dance only thirty runs for it, while a third might dance only ten, even though she, too, approves of the site.

Scouts returning from a less attractive site, meanwhile, like a hole in a stone wall, might be reporting their scores on the swarm cluster at the same time, and they could show just as much variation. Let's say these three bees dance forty-five runs, twenty-five runs, and five runs, respectively, in support of this medium-quality site. "You might think, gosh, this thing looks like a mess. Why are they doing it this way?" Tom Seeley says. "If you were relying on just one bee reporting on each site, you'd have a real problem, because one of the bees that visited the excellent site danced only ten runs, while one of the bees that visited the medium site did forty-five." That could easily mislead you.

Fortunately for the bees, their decision-making process, like that of Olympics, doesn't rely on the opinion of any single individual. Just as the scores given by the international judging committee are averaged after each skater's performance, so the bees combine their assessments through competitive recruitment. "At the individual level, it looks very noisy, but if you say, well, what's the total strength of all the bees from the excellent site, then the problem disappears," Seeley explained. Add the three scores for the tree hollow—fifty, thirty, and ten—and you get a total of ninety waggle runs. Add the scores for the hole in the wall—forty-five, twenty-five, and five—and you get seventy-five runs. That's a difference of fifteen runs, or 20 percent, between the two sites, which is more than enough for the swarm to choose wisely.

"The analogy is really quite powerful," Surowiecki says. "The bees are predicting which nest site will be best, and humans can do the same thing, even in the face of exceptionally complex decisions."

The key to such calculations, as we saw earlier, is the diversity of knowledge that individuals bring to the table, whether they're scout bees, astronauts, or members of a corporate board. The more diversity the better—meaning the more strategies for approaching problems, the better; the more sources of information about the likelihood of something taking place, the better. In fact, Scott Page, an economist at the University of Michigan, has demonstrated that, when it comes to groups solving problems or making predictions, being different is every bit as important as being smart.

"Ability and diversity enter the equation equally," he states in his book, *The Difference: How the Power of Diversity Creates Better Groups, Firms, Schools, and Societies*. "This result is not a political statement but a mathematical one, like the Pythagorean Theorem."

By diversity, Page means the many differences we each have in the way we approach the world—how we interpret situations and the tools we use to solve problems. Some of these differences come from our education and experience. Others come from our personal identity, such as our gender, age, cultural heritage, or race. But primarily he's interested in our cognitive diversity—differences in the problem-solving tools we carry around in our heads. When a group is struggling with a difficult problem, it helps if each member brings a different mix of tools to the job. That's why, increasingly, scientists collaborate on interdisciplinary teams, and why companies seek out bright employees who haven't all graduated from the same schools. "When people see a problem the same way, they're likely all to get stuck at the same solutions," Page writes. But when people with diverse problem-solving skills put their heads together, they often outperform groups of the smartest individuals. Diversity, in short, trumps ability.

The benefits of diversity are particularly evident in tasks that involve combining information, such as finding a single correct answer to a question. To show how this works, Page takes us back to the quiz show *Who Wants to Be a Millionaire?* Imagine, he writes, that a contestant has been stumped by a question about the Monkees, the pop group invented for TV who became so popular they sold more records in 1967 than the Beatles and Elvis combined. The question: *Which person from the following list was not a member of the Monkees?*

(a) Peter Tork
(b) Davy Jones
(c) Roger Noll
(d) Michael Nesmith

Let's say the studio audience this afternoon has a hundred people in it, Page proposes, and seven of them are former Monkees fans who know that Roger Noll was not a member of the group (he's actually an economist at Stanford). When asked to vote, these people choose (c). Another ten people recognize two of the names on the list as belonging to the Monkees, leaving Noll and one other name to choose from. Assuming they choose randomly between the two, that means (c) is likely to get another five votes from this group. Of the remaining audience members, fifteen recognize only one of the names, which means another five votes for (c), using the same logic. The final sixty-eight people have no clue, splitting their votes evenly among the four choices, which means another seventeen votes for (c). Add them up and you get thirty-four votes for Roger Noll. If the other names get about twenty-two votes each, as statistical laws suggest, then Noll wins—even though 93 percent of the audience is basically guessing. If the contestant follows the audience's advice, he climbs another rung on the ladder to the show's million-dollar prize.

The principle at work in this example, as Page explains, was described in the fourth century B.C. by Aristotle, who noted that a group of people can often find the answer to a puzzle if each member knows at least part of the solution. "For each individual among the many has a share of excellence and practical wisdom, and when they meet together, just as they become in a manner one man, who has many feet, and hands, and senses, so too with regard to their character and thought," Aristotle writes in *Politics*. The effect might seem magical, Page notes, but "there is no mystery here. Mistakes cancel one another out, and correct answers, like cream, rise to the surface."

This does not mean, he cautions, that diversity is a magic wand you can wave at any problem and make it go away. It's important to consider what kind of task you're facing. "If a loved one requires open-heart surgery, we do not want a collection of butchers, bakers, and candlestick makers carving open the chest cavity. We'd much prefer a trained heart surgeon, and for good reason," Page writes. Nor would we expect a committee of people who deeply hate each other to come up with productive solutions. There are limits to the magic of the math.

You have to use common sense when weighing the impact of diversity. For simple tasks, it's not really necessary (you don't need a group to add two and two). For truly difficult tasks, the group must be reasonably smart (no one expects monkeys banging on typewriters to come up with the collected works of Shakespeare). The group also must be diverse (otherwise you have nothing more to work with than the smartest expert does). And the group must be large enough and selected from a deep enough pool of individuals (to ensure that the group possesses a wide-ranging mix of skills). Satisfy all four of these criteria, Page says, and you're good to go.

Surowiecki would emphasize one point in particular: If you want a group to make good decisions, you must ensure that its members don't interact too much. Otherwise they could influence one another in counterproductive ways through imitation or intimidation—especially intimidation. "In any organization, like a team or company, people tend to pay very close attention to bosses or those with higher status," Surowiecki says. "That can be very damaging, from my perspective, because one of the great things about the wisdom of crowds, or whatever you want to call it, is that it recognizes that people may have useful things to contribute who aren't necessarily at the top. They may not be the ones everyone automatically looks to. And that goes by the wayside when people imitate those at the top too closely."

Diversity. Independence. Combinations of perspectives. These principles should sound familiar. They're versions of the lessons we learned from the honeybees: *Seek a diversity of knowledge. Encourage a friendly competition of ideas. Use an effective mechanism to narrow your choices.* What was smart for the honeybees is smart for groups of people, too.

It's not so easy, after all, to make decisions as efficiently as honeybees do. With millions of years of evolution behind them, they've fashioned an elegant system that fits their needs and abilities perfectly. If we could do as well—if we could harness our diversity to overcome our bad habits—then perhaps people wouldn't say that we're still thinking with caveman brains.

Caveman Brains

Imagine this scenario: Intelligence agencies have turned up evidence of a plot by at least three individuals to carry out a terrorist attack in Boston. Exactly what kind of attack is not known, but it might be related to a religious conference being held in the city. Possible targets include the Episcopal Church of St. Paul, Harvard's Center for World Religion, One Financial Plaza, and the Federal Reserve Bank. Security cameras at each building have captured blurry images of ten different individuals acting suspiciously during the past week, though none have been positively identified as terrorists. Intercepted e-mail between suspects appears to include simple code words, such as "crabs" for explosives and "bug dust" for diversions. Time's running out to crack the plot.

This was the fictional situation presented to fifty-one teams of college students during a CIA-funded experiment at Harvard not long ago. Each four-person team was simulating a counterterrorism task force. Their assignment: sort through the evidence to identify the terrorists, figure out what they were planning to do, and determine which building was their target. They were given an hour to complete the task.

The experiment was organized by Richard Hackman and Anita Woolley, a pair of social psychologists, with collaborators Margaret Giabosi and Stephen Kosslyn. A few weeks earlier, they'd given the students a battery of tests to find out who was good at remembering code words (verbal working memory) and who was good at identifying faces from a large set of photos (face-recognition ability), skills that tap separate functions of the brain. They used the results of these tests to assign students to teams, arranging it so that some teams had two experts (students who scored unusually high on either verbal or visual skills) and two generalists (students who scored average on both skills), and some teams had all generalists. This was important, because they wanted to find out if a team's cognitive diversity really affected its performance as strongly as did its level of skills.

The researchers had another goal. They wanted to see if a group's performance might be improved if its members took time to explicitly sort out who was good at what, put each person to work on an appropriate task—such as decoding e-mails or studying images—and then talked about the information they turned up. Would it enable them, in other words, to exploit not only their diversity of knowledge but also their diversity of abilities? To find out, they told all of the teams how each member had scored on the skills tests, but they coached only half of the teams on how to make task assignments. They left the other half on their own.

The researchers had hired a mystery writer to dream up the terrorist scenario. The solution was that a fictional anti-Semitic group was planning to spray a deadly virus in the vault at the Federal Reserve Bank where Israel stores its gold, thereby making it unavailable for months and supposedly bankrupting that nation. "We made it a little bit ridiculous because we didn't want to scare anybody," Woolley says.

Who did the best job at solving the puzzle? Not surprisingly, the most successful teams—the ones that correctly identified the target, terrorists, and plot details—were those with experts that applied their skills appropriately and actively collaborated with one another. What no one expected, however, was that the teams with experts who made little effort to coordinate their work would do so poorly. They did even worse, in fact, than teams that had no experts at all.

"We filmed all the teams and watched them several times," Woolley says. "What seems to happen is that, when two of the people are experts and two are not, there's a status thing that goes on. The two that aren't experts defer to the two that are, when in fact you really need information from all four to answer the problem correctly."

Why was this disturbing? Because that's how many analytic teams function in real life, Woolley says, whether they're composed of intelligence agents interpreting data, medical personnel making a diagnosis, or financial teams considering an investment. Smart people with special skills are often put together to make important decisions, but they're frequently left on their own to figure out how to apply those skills as a group. Because they're good at what they do, many talented people don't feel it's necessary to collaborate. They don't see themselves as a group. As a result, they often fail to make the most of their collective talents and end up making a poor decision.

"We've done a bunch of field research in the intelligence community and I can tell you that no agency, not the Defense Department, not the CIA, not the FBI, not the state police, not the Coast Guard, not drug enforcement, has everything they need to figure out what's going on," Hackman told a workshop on collective intelligence at MIT. "That means that most antiterrorism work is done by teams from multiple organizations with their own strong cultures and their own ways of doing things. And the stereotypes can be awful. You see the intelligence people looking at the people from law enforcement saying, You guys are not very smart, all you care about is your badge and your gun. We know how to do this work, okay? And the law enforcement people saying, You guys wouldn't recognize a chain of evidence if you tripped over it. All you can do is write summa cum laude essays in political science at Princeton. That's the level of stereotyping. And they don't get over it, so they flounder."

Personal prejudice is a poor guide to decision making, of course. But it's only one in a long list of biases and bad habits that routinely hinder our judgment. During the past fifty years, psychologists have identified numerous "hidden traps" that subvert good decisions, whether they're made by business executives, political leaders, or consumers at the mall. Many can be traced to the sort of mental shortcuts we use every day to manage life's challenges—the rules of thumb we apply unconsciously because our brains, unlike those of ants or bees, weren't designed to tackle problems collectively.

Consider the trap known as "anchoring," which results from our tendency to give too much weight to the first thing we hear. Suppose someone asks you the following questions:

> Is the population of Chicago greater than 3 million?
> What's your best estimate of Chicago's population?

Chances are, when you answer the second question, you'll be basing it on the first. You can't help it. That's the way your brain is hardwired. If the number in the first question was 10 million, your answer to the second one would be significantly higher. Late-night TV commercials exploit this kind of anchoring. "How much would you pay for this slicer-dicer?" the announcer asks. "A hundred dollars? Two hundred? Call now and pay only nineteen ninety-five."

Then there's the "status quo" trap, which stems from our preference not to rock the boat. All things being equal, we prefer options that keep things the way they are, even if there's no logic behind that choice. That's one reason mergers often run into trouble, according to John Hammond, Ralph Keeney, and Howard Raiffa, who described "The Hidden Traps in Decision Making" in the *Harvard Business Review*. Instead of taking swift action to restructure a company following a merger, combining departments and eliminating redundancies, many executives wait for the dust to settle, figuring they can always make adjustments later. But the longer they wait, the more difficult it becomes to change the status quo. The window of opportunity closes.

Nobody likes to admit a mistake, after all. Which leads to the "sunk-cost" trap, in which we choose courses of action that justify our earlier decisions—even if they no longer seem so brilliant. Hanging on to a stock after it has taken a nosedive may not show the best judgment. Yet many people do exactly that. In the workplace, we might avoid admitting to a blunder—hiring an incompetent person, for example—because we're afraid it will make us look bad in the eyes of our superiors. But the longer we let the problem drag on, the worse it can be for everyone.

As if these flaws weren't enough, we also ignore facts that don't support our beliefs. We overestimate our ability to make accurate predictions. We cling to inaccurate information even after it has been disproved. And we accept the most recent bit of trivia as gospel. As individuals, in short, we tend to make a lot of mistakes with even simple decisions. Throw a problem at us that involves interactions of multiple variables and you're asking for trouble.

Yet increasingly, analysts say, that's exactly what business leaders are dealing with. "Managers have long relied on their intuition to make strategic decisions in complex circumstances, but in today's competitive landscape, your gut is no longer a good enough guide," writes Eric Bonabeau, who is now chief scientist at Icosystem, a consulting company near Boston. Often managers rise to the top of their organizations because they've been able to make tough decisions in the face of uncertainty, he writes. But when you're dealing with complexity, intuition "is not only unlikely to help, it is often misleading. Human intuition, which arguably has been shaped by biological evolution to deal with the environment of hunters and gatherers, is showing its limits in a world whose dynamics are getting more complex by the minute."

We aren't very good at making difficult decisions in complex situations, in other words, because our brains haven't had time to evolve. "We have the brains of cavemen," Bonabeau says. "That's fine for problems that don't require more than a caveman's brain. But many other problems require a little more thinking."

One way to handle such problems, as we've seen, is to harness the cognitive diversity of a group. When Jeff Severts asked his prediction market to estimate the probability of the new Best Buy store opening on time, he tapped into a wide range of perspectives, and the result was an unbiased assessment of the situation. In a way, that's what most of us would hope would happen, since society counts on groups to be more reliable than individuals. That's why we have juries, committees, corporate boards, and blue-ribbon panels. But groups aren't perfect either. Unless they're carefully structured and given an appropriate task, groups don't automatically produce the best solution. As decades of research have demonstrated, groups have many bad habits of their own.

Take their tendency to ignore useful information. When a group discusses an issue, it can spend too much time going over stuff everybody already knows, and too little time considering facts or points of view known only by a few. Psychologists call this "biased sampling." Let's say your daughter's PTA is planning a fund-raiser. The president asks everybody at the meeting for ideas about what to sell. The group spends the whole time talking about cookies, because everybody knows how to make them, even though many people might have special family recipes for cupcakes, fudge, or other goodies that might be popular. Because these suggestions never come up, the group may squander its own diversity.

Many mistakes made by groups can be traced to rushing a decision. Instead of taking time to put together a full range of options, a group may settle on a choice prematurely, then spend time searching for evidence to support that choice. Perhaps the most notorious example of rushing a decision is a phenomenon that psychologist Irving Janis described as groupthink, in which a tightly knit team blunders into a fiasco through a combination of unfortunate traits, including a domineering leader, a lack of diversity among team members, a disregard of outside information, and a high level of stress. Such teams develop an unrealistic sense of confidence about their decision making and a false sense of consensus. Outside opinions are dismissed. Dissension is perceived as disloyalty. Janis was thinking, in particular, of John F. Kennedy's reckless decision to back the Bay of Pigs invasion of Cuba in 1961, when historians say that President Kennedy and a small circle of advisors acted in isolation without serious analysis or debate. As a result, when some twelve hundred Cuban exiles landed on the southern coast of the island, they were promptly defeated by the Cuban army and tossed into jail.

Decisions made by groups, in short, can be as dysfunctional as those made by individuals. But they don't have to be, as the swarm bees have already shown us. When groups contain the right mix of individuals and are carefully structured, they can compensate for mistakes by pooling together a greater diversity of knowledge and skills than any of their members could obtain on their own. That was the lesson of the experiments Hackman and Woolley conducted in Boston: Students did better at identifying the terrorists when they sorted out the skills of each team member and gave everyone a chance to contribute information and opinions to the process. Simply by drawing from a wider range of experiences, as Scott Page's theorems proved, groups can put together a bigger bag of tricks for problem solving. And when it comes to making predictions, like how many gift cards will be purchased this month, groups can cancel out personal biases and bad habits by combining information and attitudes into a reliable group judgment.

What sets the bees apart, in this respect, is that they don't seem to be fighting their instinctive nature as much as we are. As individuals, bees may show a lot of variety in the way they go about their tasks, but when it comes to the swarm's goals, they share a singularity of purpose. How do they make it look so easy?

Part of the answer has to do with the fact that honeybees in a swarm are more closely related genetically than any human group you can name—even a big family reunion. They're all sisters, in fact—or rather, half sisters, since their common mother, the queen, was fertilized by a number of males (who live only a few days). That means the bees have a powerful stake in the welfare of the group, so powerful that they often put the group's interests ahead of their own.

But that's only part of the answer. Even if the bees are eager to cooperate with one another, as they're genetically programmed to do, that doesn't explain *how* they do it so well. Natural selection may favor the fitness of honeybees at the colony level, just as it does for Deborah Gordon's desert ants, but what enables them to act as one when making decisions?

"It's actually a brilliant approach," says Kevin Passino, a professor of engineering at Ohio State who has collaborated with Tom Seeley and Kirk Visscher on their honeybee house-hunting experiments. Using data from Appledore Island, Passino has created computer simulations of swarms that show a striking resemblance between their collective decision-making system and the mental processes of a single individual. "If you think about the way the system works," he says, "the swarm is *paying attention* to its surroundings as if it were a single brain."

Just as a brain processes information about the outside world through countless neurons, the swarm collects and interprets information through the scout bees. When they perform waggle dances on the surface of the cluster, the scouts are not only sharing with one another information about what's out there, they're also giving the group as a whole a broad "field of view." You can see it for yourself if you take a close look at the cluster surface, Passino says. When more bees are dancing in one direction than another, that's a physical representation of which site is ahead in the competition. If you look closely at the swarm, you can actually *see* what it's thinking.

"Just as humans are very good at looking at a number of different objects in a scene and finding the one we're interested in, so are the swarms," Passino says. "Our brains use parallel processing, and so do the swarms. They can look at a whole mess of things in a scene, dismiss the ones that aren't what they're looking for, and find the one they want. Because they're paying attention to the right thing."

In a sense, the behavior of scouts on the swarm cluster and their distribution among the different nest sites is the equivalent of a "group memory," Passino explained. "It's not a long-term memory, like those you might have of a summer vacation, but a very short-term memory—more like a holding of information, or of preferences, that the swarm builds on and uses. It's a spatial mapping of information from the external world onto the swarm. The exact same thing happens with a brain."

As a group, the bees exploit their diversity of knowledge through a decentralized process that lets them see their choices clearly. They don't rush to select the first nest site that meets their criteria, but rather take enough time to sort out all the options, allowing the cream to rise to the surface. None of the bees takes it upon herself to decide for the whole group. None even knows what all the others are doing. In the end, Passino says, "what the swarm knows is actually far more than the sum of what is known by the individual bees, since it includes the information stored in the bees' brains *and the information coded in the locations of the bees and their actions.*"

To deal with an urgent problem, in short, the bees take advantage of the swarm's own complex structure to pursue a common strategic goal—a lesson that wasn't lost on one determined business leader when he found himself in the hot seat.

The Beer Game

Dennis O'Donoghue was feeling upbeat as he stood before fifty of his fellow test pilots, engineers, and managers not long ago in the auditorium of Boeing's 2-22 Building in Seattle. As a vice president at Boeing, he told them how happy he was that they'd gathered that morning to play what he described as the Beer Game. Contrary to the way it sounded, the game didn't involve drinking, he said. It was actually a role-playing simulation of a beer distribution system. Its purpose was to give players a taste of how difficult it can be to make good decisions when you're part of a complex system. O'Donoghue wanted everybody in the room to understand this, he said, because he needed their help to tap into the problem-solving wisdom of his 3,600-person division, which he believed was currently being squandered.

O'Donoghue's group, known as Flight, Operations, Test & Validation (FOT&V) was about to become the main focus of the 787 Dreamliner project, possibly the most ambitious endeavor Boeing had ever attempted. Boeing had sold more than eight hundred 787s worth about $170 billion to more than fifty airlines. But before a single plane could be delivered to a customer, it had to be thoroughly tested and certified by mechanics, engineers, and test pilots from O'Donoghue's team. At the time of the Beer Game exercise, the 787 program had already slipped fourteen months behind schedule, and FOT&V was still waiting for the first plane to test.

Boeing was taking a huge gamble with the Dreamliner, which the media had hailed as a "game-changing" innovation. About half of the plane consisted of lightweight composite materials such as carbon-fiber reinforced plastic, which was common on military aircraft but had never been used so extensively on a commercial plane. To produce the 787s faster, and to spread out the hefty financial risk, Boeing had also committed to a radically new manufacturing process that relied on scores of suppliers around the world. So far the gamble had paid off. Boeing's customers had found the 787's design irresistible—a speedy, midsize plane that could fly as far as ten thousand miles on 20 percent less fuel—setting industry records for sales.

Since FOT&V would be the last unit in the production process, O'Donoghue and his team knew that they would be under intense pressure to meet their deadlines, no matter what complications they encountered. Otherwise, Boeing could face financial penalties from customers, which could total billions of dollars. But that wasn't going to be easy, since delays earlier in manufacturing meant that the test flight schedule had been severely tightened. Instead of getting nine to fifteen months to validate the 787, as they'd had with the 777, FOT&V would get only six and a half months. And to make sure everything got done, they'd be working with six to nine planes at the same time, instead of four as before, which meant juggling more maintenance crews, pilots, technicians, and parts than ever. On top of that they'd have their normal load of regular production aircraft—747s and other models—to test before they were delivered to customers.

So here was an organization facing a critical task involving multiple, simultaneous interactions among several thousand individuals, under considerable stress, with a brand-new leader. And what did that leader do? He decided to gamble a chunk of their precious time to figure out how to tap into the collective intelligence of the group—their *diversity of knowledge*—to get them through the crisis.

The key to O'Donoghue's strategy came from a small team of Boeing analysts who showed him that FOT&V, like an ant colony or beehive, was a complex adaptive system that had evolved to perform specific tasks very well. The complexity, in this case, came from the number of variables that FOT&V had to cope with, many of which were affected by other variables—thousands of tests on multiple airplanes on the ground, in the air, and in wind tunnels, planned and unplanned delays called layups, with a dozen departments sharing resources. Such a system, they told O'Donoghue, can be difficult to manage, let alone change, unless you understand how it works. You have to recognize, for example, that a decision made in one part of the system can affect those in other parts in ways that are hard to predict. Under such circumstances, you might think you're moving briskly toward your goal when you get blindsided by an unpredictable delay. All of which convinced O'Donoghue and his team that they needed to rethink everything FOT&V did before it was too late. "Just for sheer magnitude of change, this was off the scale, compared to anything we've ever done before," says one of the Boeing consultants.

The Beer Game was part of this process of rethinking. Here's how it was played: There were four persons on each team. One represented the Retailer. "You're the owner of a 7-Eleven on a corner in Carnation, Washington," O'Donoghue says. "You sell beer, gas, potato chips, eggs, milk, and a bunch of other stuff. And one of the brands you sell is Dreamliner Beer, which comes from a small brewery in the Pacific Northwest. It's not your most popular beer, but you consistently sell a few cases a week."

Another player was the Wholesaler. "You drive the truck that delivers the beer," O'Donoghue says. "You don't deliver just beer. You also deliver milk, soda, and other beverages." The third player was the Distributor. "You own the warehouse and handle lots of kinds of beer, not just Dreamliner," O'Donoghue says. The fourth player was the brewer who manufactured Dreamliner beer.

The game was played in rounds that each represented a week. In the first week, players took orders for Dreamliner beer from the player to their left and placed new orders with the player to their right. Because of the time it took to process and ship these orders, there was a four-week delay between the time a player placed an order and when it was delivered. (This turned out to be a critical factor, as everyone later discovered.) The goal of the game was to keep the beer moving as efficiently as possible through the system. Too much inventory wasn't good. Not being able to fill an order was worse.

"The last thing you want is customers getting mad at you," O'Donoghue says. "Like that big guy in the Ford truck who pulls up to the pump at your 7-Eleven and says, 'Dude, what do you mean you're out of Dreamliner beer? Forget about me filling up my F-350!'"

I'd been invited to play the game, too. My teammates included Vern Jeremica, chief pilot of flight standards, who was given the role of Retailer. Our Wholesaler was Andy Deutsch, an industrial engineer with the test manufacturing division. Leon Robert III was our Distributor. Leon was a pilot who had just flown a 767 in from Singapore to be reconfigured as a freighter. That left the Brewery spot to me. O'Donoghue had told us up front that the Beer Game was about teamwork. "We're all in this together," he'd said. "Everything you do has an impact on somebody else." But we didn't really understand what he meant until we were well into the playing of it.

Vern started us off by picking up a card from a deck that represented customer orders. He studied the card, then placed it facedown in a discard pile. Throughout the game, only he would know how many cases of beer customers at his 7-Eleven had purchased that week. The rest of us would have to guess. Then he wrote a number on a piece of paper, which represented his order, and put it facedown on the table next to Andy.

While Vern was doing that, Andy picked up a piece of paper that was supposed to be Vern's previous order. It was for four cases of beer. Andy filled that order by pushing four poker chips into a box that represented his truck. Then he wrote his own order for four cases and put it facedown on the table next to Leon.

Leon, meanwhile, was doing pretty much the same thing in his role as Distributor. After looking at Andy's previous order for four cases, he pushed four chips out of his warehouse. Then he wrote an order for four cases and put it down in front of me. My job was slightly different, since I produced the beer. But the moves were the same. After looking at the order from Leon, I pushed four chips into a box that represented manufacturing, then I wrote a production order for four more cases at my brewery.

That was Week 1.

The next three weeks were exactly the same, as we got used to the rhythm of receiving shipments, filling orders, and placing new orders—always for four cases. We'd each started the game with twelve cases in our inventories, and that's how many we still had. Four cases came in each week, and four cases went out. Then, in Week 5, something different: Vern picked up a card that said his customers wanted eight cases. "Somebody must be having a party," he said to himself. In Week 6, he sold another eight cases, and in Week 7 he sold twelve. "Here we go," he said. "I'm getting cleaned out."

When Andy got Vern's order for ten cases in Week 8, he bumped his order to ten cases too. His inventory was almost empty. He needed to catch up.

When Leon got Andy's bigger order, he resisted the urge to panic. He'd expected ups and downs in customer demand. So Leon ordered only six from me. "Just like the flood, there has to be a drought coming," he figured. But by Week 13, he was cleaned out, and by Week 16 he had a backlog of orders for twenty-five cases.

I didn't feel the spike in orders that was traveling through our system until Week 18, when I got a slip from Leon for fifteen cases. I was deliberately trying not to produce too much beer, but my cautious approach was hurting Leon, who fell further and further behind. "I'm going to get a new supplier," he grumbled in Week 20, when his backlog ballooned to fifty-four cases.

Vern's customers, remember, had increased their orders at the 7-Eleven by only a few cases, and now Leon had a backlog of fifty-four cases. How did our system get so out of whack? It wasn't because of Vern's customers, despite what we suspected. As we later learned, after the brief surge to twelve cases, their orders had stayed at eight cases for the rest of the game. They never changed again. Somehow that initial ripple had turned into a tidal wave. By Week 23, Andy was ordering fifty cases a week, and soon after that so was Leon. As the manufacturer, I was also overshooting on production. Our supply chain was out of control.

As far as we could tell, none of us had overreacted. At each step, we'd resisted the urge to request extra beer until we got those alarming orders from our teammates. But our efforts to manage the surge had been useless. We'd had no success at all in moderating demand. In fact, the only thing we'd accomplished was to create massive backlogs in our inventories. At one point, Leon had fallen behind by 304 cases.

That was roughly the moment when the wave of orders moving from Retailer to Brewer hit the wall at my end of the distribution line and bounced back toward Vern with a vengeance. Now Leon and Andy were receiving 50 cases of beer a week, even though they hadn't ordered a single case for more than a month. By the end of the game, Andy was drowning in 163 cases of Dreamliner beer. We had made a real mess of it.

"So how did that make you feel?" O'Donoghue asked with a smile when the teams had finished. "Did you get frustrated? Did the game make you feel helpless? Did you want to blame somebody else for your problems?" Don't take it too hard, he said. The same thing happens to everybody who plays the Beer Game, from high school students to CEOs of major corporations, and it's been happening since the 1950s, when the game was invented at the MIT Sloan School of Management.

What causes players to make these mistakes? Why do they

lose control so quickly? The answer has to do with their perception of cause and effect. The four-week lag time between ordering and receiving beer is simply too much for most players to grasp. Their caveman brains can't adjust. Social scientists describe this as a hidden trap called "reductive bias," in which a person treats a complex phenomenon as if it were a simple one. But who can blame them, in this case? If you were driving down the road and had to cope with a four-second delay between the moment you turned the steering wheel of your car and the moment the car changed direction, you'd wreck it every time, too.

"It was pretty easy to get into trouble, wasn't it?" O'Donoghue asked. "Well, imagine if your company had to deal with fifty wholesalers instead of one." Businesses get swept up in boom-and-bust cycles all the time, he says. And then what happens? People get fired. People get laid off. Factories close down. Yet it's the dynamics of the system that causes the instability.

O'Donoghue didn't have to mention it, but Boeing's 787 program had become a prime example of this. Delays by key suppliers in completing work had brought construction of the first few planes to a grinding halt at the main plant in Everett, Washington. That had forced Boeing to devote extra resources to finishing the work, which contributed to even more delays.

Like many others who have played the game, our team was done in by a kind of quiet panic. In a desperate attempt to make up for continuing shortages in our inventories, we kept ordering beer week after week. What we failed to recognize was that the pipeline was already full, and that by trying to "manage" the situation we were just making it worse. By focusing on our own portion of the supply chain, we magnified the instability of the system as a whole, which blew up in our faces.

That kind of instability was exactly what Dennis O'Donoghue wanted to avoid in FOT&V. If he and his team were to weather the coming Dreamliner storm, they had to understand the complexity of their system as a whole. But before his organization could gain that broader "field of view," O'Donoghue and his team had to untangle FOT&V's underlying social structure—one that celebrated freewheeling heroes and fierce warring tribes.

Firefighters and Clan Chieftains

The one thing Dennis O'Donoghue knew for sure as the Dreamliner program bore down on his organization was that he didn't have any time to lose. Unless he found a way to dramatically speed up the pace of flight test operations, the volume and complexity of the tasks about to descend on FOT&V would be crushing. Business as usual wasn't going to make it.

In the old days, when the testing schedule included more wiggle room, teams of pilots, engineers, mechanics, and other specialists were assigned to a single test airplane from start to finish. If they had a problem with their plane, they handled it. If they needed a part, they got it. If they wanted more runway time, they worked it out with other crews. Each decision was ad hoc.

For forty years, that's how flight testing was done at Boeing. But it wasn't very efficient. When O'Donoghue took over, teams were meeting their deadlines only about half of the time.

One of the main reasons they tended to be late was that they were constantly dealing with crises. The right equipment wasn't available. A test program was changed, and no one told the maintenance crew. Many times, problems had nothing to do with the airworthiness of the aircraft, which was the focus of their work. They were unintentionally created by the teams themselves. "FOT&V had evolved into a large, complex system," O'Donoghue says. "But many people didn't see themselves as part of a system. They saw themselves as firefighters working in isolation. They would put out one fire after another, making short-term fixes but never really addressing fundamentals. And that just caused other brushfires down the road that somebody else had to go fight."

If engineers on a test flight, for example, discovered that certain monitoring gear hadn't functioned correctly during a test, but the weather and other conditions were still good for flying, they might decide on their own to repeat the test, even though that meant keeping the plane in the air a few hours longer than planned. No big deal, right? They were just taking care of business. But on the ground a maintenance team was waiting to reconfigure the aircraft for another test the next day, switching out instrument packages, repositioning ballast, and so on. So the delay put them behind schedule. Which meant the plane wasn't ready when a new flight crew showed up in the morning. And when the crew finally rolled the plane out on the runway, the fuel truck that was supposed to be there had long ago departed to service other aircraft. The impact of the engineer's decision rippled through the system.

That was the lesson of the Beer Game, of course. Players who make quick fixes at one position in a complex supply chain often cause other problems downstream. "But that's a hard thing for people to understand," O'Donoghue says. "For a long time in our organization, everybody has wanted to be the go-to person, the one who saves the day. And we've rewarded those people. We've actually developed a hero worship of firefighters." That wasn't what FOT&V needed now, he says. Now it needed people who realized they were part of a complex system, people who could meet deadlines under pressure—not 50 percent of the time, but 90 percent of the time. By the same token, he needed a system whose structure was flexible enough to tap into the problem-solving abilities of the group.

So far, he didn't have either.

To help straighten out his division's outmoded habits, O'Donoghue turned to his team of systems analysts, some of whom had come to him from Boeing's strategic planning unit called the Phantom Works. Their assignment was to map out the way FOT&V did business and identify leverage points for improvement. They began by creating a chart of the organization's work flow.

"We'd never modeled it before," O'Donoghue says. "And it was really fascinating, because people stood there in front of the chart as we started to draw it and the conversations about what did what to whom were really interesting. There were a lot of misconceptions. A lot of disagreement about how the system really worked."

What emerged was a realization that what really made things tick in FOT&V had less to do with formal responsibilities than with informal networks of alliances and loyalties. "That was the big surprise," says Bob Wiebe, one of the analysts. "We discovered these powerful informal networks that were basically tribal, just like the Scottish highlands or something. If you had a problem on the job, you'd call someone in your trusted network to find out what could be done about it. And before you went forward, you'd get the nod from the clan chieftain. Loyalty to the clan was the way you did business."

The two biggest clans, it turned out, were the maintenance and engineering groups, who'd been feuding like the Hatfields and McCoys as long as anyone could remember. "Nobody knew why. It had just always been that way," says Karen Helmer, another Boeing analyst. One explanation for the friction emerged when the consultants mapped out various roles played by individuals in each clan. Many overlapped. Different clans, in other words, were performing the same functions, which was a hidden source of the rivalry.

It hadn't helped that FOT&V had been patched together decades ago by merging several clans. "Groups like flight test engineering, test manufacturing, flight operations, quality, all came from different parent organizations," O'Donoghue says. "They were put in this organization, but they came with their own cultures, their own processes, even their own tools. They'd never really been integrated, and consequently they worked against each other."

What they needed, he concluded, was a common sense of purpose. "The paradigm I like to use is that the organization, before I took over, was a complex adaptive organism that lacked a central nervous system. The different parts had adapted on their own, based on what they perceived the threats to be, and, interestingly, in some cases, they perceived other parts of the organization as threats, and so they worked against them. What I wanted to do was create a central nervous system that would provide strategic direction to them all."

During his time as a test pilot in the Marine Corps, O'Donoghue had been stationed at Patuxent River Naval Air Station, where he'd seen a maintenance operations center that managed as many as forty test flights a day, coordinating countless tasks to prepare, maintain, and configure each aircraft. When he took over FOT&V, he knew he wanted that kind of capability. "We're going to have our own test operations center," he told his management team. "Let's go figure how to do it."

The main function of the Test Operations Center (TOC), as he imagined it, would be to give the organization better vision. By tracking the status of every airplane in the test fleet, as well as every resource required to conduct every flight test, from crews to runways, the TOC would provide the "fleet perspective" the organization needed to avoid tripping over itself, as it used to do in the past. Above and beyond that, it would serve as a vehicle to carry out collective problem-solving. When crews in the field ran into problems they couldn't handle within a reasonable time-frame, the TOC would convene a group of experts from various parts of the organization to come up with potential solutions. Like scout bees searching the neighborhood for a solution to the swarm's problem, these individuals would gather the best information and advice available from their areas, whether they were maintenance, test integration, design, or flight crews, and put it all on the table for consideration.

"What it all really comes back to is the wisdom of the crowd," O'Donoghue says. "I can't possibly know and control everything that goes on in this organization of 3,600 people. Top-down control does nothing but kill the ability to adapt and innovate. So we're trying to preserve the organizational intelligence we already have." The idea, he says, was for the TOC to serve as a clearinghouse for information from all over FOT&V. Rather than preempting local problem-solvers, it would tap into their expertise during a crisis. "We have to push the decision making down to the appropriate level where somebody has the answer."

Let's say a flight crew smells smoke in a test plane while taxiing toward a runway. After notifying the TOC of the situation, they conclude it's not an emergency. Still, it wasn't something they'd expected. So they put the day's tests on hold to do some investigating. But after fifteen minutes of going through checklists, they're no closer to finding the cause of the smoke than before. They don't want to risk missing their deadline to complete the package of tests, so they turn to the TOC for help.

The Test Operations Center is a 32,000-square-foot complex of cubicles on the fifth floor of the FOT&V building in Seattle.

At its heart is a 2,000-square-foot control room manned twenty-four hours a day by a dozen or so specialists from all over the organization, including engineers, mechanics, and other experts. Their job is to monitor the status of the test fleet, to make sure everything's in place when it's needed, from parts to tools to flight crews, and to anticipate conflicts or problems—in short, to determine what to do with each plane in the fleet on a daily basis. Just as the scout bees created a short-term group memory on the swarm cluster by recording information about the neighborhood through waggle dances, so the TOC helps keep the organization on track by paying attention to the bigger picture. A six-by-sixteen-foot flat-panel screen on one wall of the control room shows a half-dozen displays of the fleet's status in real time.

"When the call comes in, it's assigned to a specialist who owns that problem until it's solved," says Janet Mueller, who manages the TOC. "If it's a maintenance problem, the owner is likely to be a senior maintenance person. Within about a half-hour, that person pulls together by telephone a group of subject matter experts from all over the organization to come up with recommendations. Some of them might not even be in this country. We have partners around the world. They come up with all of the possible solutions. How long's it going to take? What are the skills? Do we have the parts? Do we have the paper? Do we have the tools to do it? It's up to the team to decide what to do and agree on a plan."

If the cause of the smoke in the test plane turns out to be a noncritical problem—like a faulty electrical part in the galley, for example—then the ops center team might decide not to delay the rest of the test package, which has nothing to do with the electrical system. That would be good news for the flight crew, which is eager to get finished. But if the smoke was coming from electrical instruments necessary to test the plane, and there were no replacements available today, that might change the crew's plans. Then another specialist in the control room, called the "fleet optimizer," would get into the act. Her job is to consider how changes to one plane's testing program could have a cascading effect on the whole fleet. Should the plane be reassigned to another crew? Should the test package be assigned to another plane? How will the delay affect the next test package?

In the past, such problems were solved locally without considering the ripple effect on the fleet. But with six planes doing tests at the same time, the chances of unintended consequences were too great. "What's happening is, we're looking ahead. We're asking questions. We're interacting with people we haven't traditionally worked with in a face-to-face, eyeball-to-eyeball type of situation," says Mueller. The result has been that TOC has helped the organization as a whole handle the workload more efficiently. "It's helping us to say, okay, we need to look at this a little better, maybe change the process flow, and make sure it makes sense."

The lesson of the TOC, in other words, was that when you're managing a complex system in which many tasks can be interrupted or delayed by many others, and cause and effect can be difficult to figure out, the best you can do to stay on schedule is to identify leverage points where small changes can lead to big fixes. In that sense, O'Donoghue and his colleagues were inventing a system that would enable FOT&V to adapt to changes in the environment as rapidly as Deborah Gordon's ant colony had done in the desert, but to do so under the guidance of a small team of problem solvers selected from the group as a whole. Such teams, they hoped, would make the most of their different backgrounds, abilities, and perspectives (as well as, literally, their different toolboxes) to come up with creative solutions.

What they didn't anticipate at that point were yet more delays for the 787. Boeing had originally promised delivery of the first Dreamliner to All Nippon Airlines in Tokyo by May 2008. But the first test plane wasn't turned over to O'Donoghue and his team until late May 2009. When structural problems turned up where the plane's wing joined its fuselage, the testing and production schedule was revised again, pushing the delivery into late 2010 or early 2011. At the same time, Boeing decided to expand FOT&V's workload by giving O'Donoghue's division responsibility not only for all commercial airplanes but also for military aircraft, including tankers, airlifters, fighters and rotorcraft, as well as space and missile defense. The new organization would be called Boeing Test and Evaluation. "It just made sense to integrate all these activities that were going on at twenty-eight different sites in the enterprise," O'Donoghue says. "So my organization pretty much doubled in size overnight."

Meanwhile, activities went forward in the TOC as teams tested a new carbon brake system for the 737, evaluated the Navy's P-8, the Poseidon, which is a maritime patrol version of the 737, and tested a freighter version of the 777, with a new model of the 747 waiting in the wings. "The problem solving is going on twenty-four/seven," O'Donoghue says. "We had a meeting this morning at one o'clock."

How were the Boeing teams able to keep up? How could they make the right calls at the right moment? Part of the answer has to do with an experiment conducted a few years ago by Patrick Laughlin, a social psychologist at the University of Illinois at Urbana-Champaign. The experiment was designed to assess the abilities of small problem-solving groups like the ones convened by the TOC. It focused on the kind of task that has a single correct answer (such as, where's the smoke coming from?) rather than the kind that depends on somebody's opinion (such as, who should be the next American Idol?). Laughlin called the first kind an "intellective" task and the second a "judgmental" task, and he wanted to know whether small groups might be as successful at intellective tasks as individual experts.

To find out, Laughlin and several collaborators recruited 760 students at the university and asked them to solve two difficult problems involving a series of ten letters. The letters had been coded using the numbers 0 through 9, and the objective was to figure out which letters stood for which numbers. Two hundred of the students were asked to work on the problems by themselves, and the rest were divided up into groups of two to five persons. Everybody was allowed up to ten guesses, with groups deliberating and reaching a consensus among their members before making each guess.

The results were surprising, even to Laughlin. They showed that groups of three or more people not only performed better than the average individual, they also did better than the *best* individual. That meant groups of three or more did better than even the smartest people in the groups, which Laughlin called "a striking and unusual result." What was equally impressive, considering what we've learned about smart decision-making by hundreds of honeybees, was how few people were needed in each group. With as few as three people, the experiment proved, a group possessed enough diversity of knowledge and problem-solving skills to move to the front of the class.

This was only possible, Laughlin cautioned, because of the specific nature of intellective tasks. When a group tackles a problem of this kind, such as a mathematical or logical puzzle, he says, its members can build upon each other's insights, abilities, and strategies to improve one solution after another, until one of them cracks the problem. That was good news, in a sense, not only for Janet Mueller's problem solvers in the TOC, but also for medical personnel searching for the cause of a patient's distress, detectives solving a crime, or scientists on a research team. As long as members of such groups actively cooperate with one another, deliberately building on each other's strengths—as Richard Hackman and Anita Woolley discovered with their antiterrorism teams—they can compensate for many of the hidden traps of our caveman brains, and make good decisions.

The same wasn't true, necessarily, for groups tackling problems of the other kind, judgmental tasks, where the objective wasn't to find a single correct answer but to achieve consensus on issues based on values as well as reasoning—such as whether smoking should be allowed in restaurants, which candidate would make the best mayor, or whether individuals of the same sex should be allowed to marry. To tackle questions like these, a group might need to take an entirely different approach—one that taps diversity of knowledge just as diligently, but that looks as messy as swarming bees.

Just ask the citizens of Vermont.

Town Meeting Day

They looked friendly enough as they entered the auditorium of the old academy building in Bradford, Vermont—men, women, and children in a sea of flannel, denim, and fleece. Neighbor greeted neighbor. Mothers showed off babies. Volunteer firemen joked as they stood against the back wall. But make no mistake: this was a crowd on a mission. The citizens of Bradford were determined to take control of their town's spending, and this was their chance.

It was the first Tuesday after the first Monday in March, and all across Vermont town meetings were being held to elect local officials, discuss community issues, and make decisions as a group. Should they renovate the police station? Bring broadband Internet to town? Buy a new fire truck? Pave muddy roads? These weren't public hearings taking place. They were assemblies of citizen legislators. Townspeople weren't coming to be advised or consulted. They were coming to pass laws. They've been doing it this way for more than three centuries in New England, where residents practice what is arguably the most authentic form of democracy in the world.

Like the citizens of ancient Athens, people in Vermont towns have reserved the right to make local laws for themselves. More than eighty communities still do it the old-fashioned way through something called a floor meeting, in which binding decisions are made through face-to-face deliberations. Those who champion this form of decision making say it's well worth the extra time and effort, not only because it's more efficient but also because it teaches tolerance for the opinions of others. By promoting a sense of belonging, they say, it lends an invisible strength to the community as a whole.

Because of the way they reach decisions, town meetings also bear an intriguing resemblance to what biologist Tom Seeley called the "animal democracy" of honeybees. Just as scout bees report on the surrounding countryside, so townspeople bring to meetings a wide range of facts and opinions about their community. Just as bees recruit one another to support their preferences, so townspeople engage in civil but often spirited debates. And just as a swarm chooses a new nest site by establishing a quorum of supporters, so townspeople resolve their differences by voting—shouting aye or nay, standing, or using handwritten ballots. "In both town meeting and the swarm bees, the heart of the decision making comes down to a competition of ideas among individuals," with each individual making his own decision independently, says Seeley, who attends town meetings in Maine, where he owns a home. *Seek a diversity of knowledge. Encourage a friendly competition of ideas. Use an effective mechanism to narrow your choices.* What works for the bees seems to work for town meetings too—and the reason has to do with the local nature of the problems they address.

The first town meeting in Bradford took place in 1773, and the issues that early settlers dealt with weren't so different from today's. They concerned budgets, road maintenance, and management of common property. At one meeting, it was decided that pigs could run in the highway as long as they were properly yoked, but that stray cattle or horses would be impounded. When Jesse Woodward requested compensation for a horse that died after falling through a public bridge in disrepair, the town refused—showing a thrifty streak even then—though it agreed, for some reason, to pick up May McKillip's surgeon's bill when she broke her arm on the same bridge. Such issues may seem inconsequential compared to the revolutionary ideas about life, liberty, and the pursuit of happiness that were being debated in Philadelphia at about the same time. Yet the principles at stake were important—how should we govern ourselves? How should we reach decisions about matters that directly affect us when we strongly disagree with one another? What should the mechanisms be by which we make good decisions as a community?

These were the issues being played out in the Bradford auditorium, where some 70 people had gathered by nine a.m. Outside, snow was piled knee-high along the walkways, and more flurries were forecast. If today's meeting followed the usual pattern, about 10 percent of the town's 2,600 people would show up. Many would be senior citizens, who could afford to devote a full day to the meeting, and all but a handful would be white, like the community as a whole. These days, the rising cost of everything was topic number one in Bradford households, where incomes tend to fall below the national average. That was one reason, no doubt, why the town budget was today's main target.

Larry Coffin, a retired social studies teacher, rapped the podium twice with a small gavel. "Call together the Bradford town meeting," he announced. This was Coffin's thirty-seventh year as moderator, a role that put him in a unique position to shape the flow of the day's events. Part coach, part referee, his job was to run the meeting as fairly and efficiently as possible, following state law and Robert's Rules of Order. That meant guiding Bradford's citizens through the confusing process of motions and amendments, while not getting in their way. This day, Coffin believed, belonged to the people.

He began in the traditional manner by asking for a moment of silence to remember friends and neighbors who had passed away this year, to recognize those serving their country, and "out of respect for the exercise in democracy that we are about to engage in." A trio of Boy Scouts led the group in the Pledge of Allegiance, and Coffin reminded the group that the Methodist Church next door would host a benefit lunch about noon. After briefly outlining the day's agenda, he reminded everyone to address their comments directly to him, rather than to other citizens, to keep debate from getting personal. On matters of procedure, his decision would be final, he said. "But if at any time you feel I have made a mistake, please speak up and say so, and I will not leave the stage in tears."

The business of the meeting was to vote on fifteen questions, or articles, which had been advertised, or "warned," by town officials a month ago. Apart from the election of new town officials, the most important business of the day was article 4: "To see what sum of money the Town will vote for current expenses for the year 2008, and to vote time and manner of collecting the same." The assembled group, in other words, was being asked to fix the total amount for the coming year's budget.

One person in the crowd had been waiting for this moment. His name was Larry Drew, and every year at this point in the meeting he would raise his hand to be recognized. And every year he would make the same motion—to cut the budget down to size. This year was no different. "We have a poor community," he said. "A lot of people can't pay their bills. So I champion the taxpayer, I guess."

A grandfatherly figure in a blue plaid shirt, Drew stood and read a long motion full of legalese that ended with the number $628,425, which was $56,494 less than what the select board had recommended in their budget. "I'm willing to defend this budget if you're willing to hear. Or I can just take my seat, if I'm going to get shot down again like in previous years," he said.

That got a laugh from the crowd, which encouraged Drew to keep going. "We need to be a little more conservative," he said, taking a sharpened pencil from behind his ear. For the next ten minutes he went through the budget line by line. Why spend $2,000 on office supplies when you can get by on $1,500? Why buy fancy new high-band radios for the police force if the old ones are just fine? "If Homeland Security is mandating it, let them pay for it," he said. And what's this about a town program to encourage energy conservation? "Somebody's going to tell me I can't turn on a lightbulb? I think that's kind of silly." By the time he was finished, Drew had people smiling and nodding their heads. Or at least smiling, which was good.

"Now you can fire at will," he said, sitting down.

"Well, I think you've managed to gore everybody's ox, Larry, so let's move along," Coffin remarked, getting a laugh himself.

Then something remarkable happened. Another town elder named Dan Perry, who was not known to be one of Drew's allies, rose to speak. "This may be the first time in thirty-five years I have ever agreed with Larry," he said, "but I think he's right. We're all taking steps in our own lives to adjust, and as a town we need to also."

"We can't hear you," someone in the back called out. "Say it again."

"I don't know if I *want* to say the whole thing again," Perry grumbled.

"Okay, next time then," someone joked.

Now that Drew's motion was on the floor, something had to be done about it. The group could amend it or put it to a vote. "I want to amend his motion to $656,000," Todd Terrill said, splitting the difference between Drew's proposal and the select board's budget. The motion was seconded.

"I want to amend Todd's amendment," Peggy Coutermarsh said, "to $590,634." That was the same amount as last year's budget.

Things were getting complicated fast. Coffin called for discussion on Coutermarsh's motion, and one of the selectmen, Robert

Miller, offered a belated defense of the board's budget. "We worked hard on that budget and I would suggest we stick to it," he said. But that horse had already left the barn. The question on the table wasn't whether to cut the budget, but by how much.

"I'm a little confused now," said a man with a billy-goat beard. "If we vote yes for Peggy's motion, will it wipe out Todd's?"

"No, it won't," Coffin said. "We'd deal with Todd's next."

"Well, I suspect that nobody knows it as well as you, Larry," said Douglas Miller, Robert's father, and a former selectman himself. "I think we best let you lead us along by the hand."

So Coffin did just that, calling for a vote on the secondary amendment first. "All those in favor of Peggy's amendment?" A handful of people said aye. "All those opposed?" The crowd boomed nay. "The nays appear to have it," Coffin said.

Next came Terrill's amendment, which received a much closer vote. "The nays appear to have it," Coffin said again. But this time his opinion was challenged.

"Division," someone shouted. That meant a standing vote would be required (hence the phrase "Stand up and be counted").

"All those in favor of Todd's amendment, please stand," Coffin asked. "Now count off, please, beginning in the front of each section." By the time the last person called out his number, those in favor had forty votes. The nays totaled eighty-nine votes. "The motion to amend has failed," Coffin said.

Sensing a chance to rescue their budget, David Chipman, the select board's chair, suggested that the meeting take a break for lunch so the board could reconsider their position. "Do I have a motion to recess?" Coffin asked. He got one.

"All in favor?" Aye, said the members of the select board.

"Opposed?" Nay, boomed the rest of the town.

It was almost noon, and the crowd seemed to be losing steam. A new amendment was offered by David Allen to substitute $684,919 for the figure Larry Drew had offered. That was the amount of the select board's original budget.

"If we do that we'd be raising the budget by nearly a hundred thousand dollars over last year," someone called out. "Where's all the money going?"

"It's called a budget," said one of the selectmen testily. A man named Jerry Martin stood up and tried to get the discussion back on track. "It looks like the majority of people would like to see cuts," he ventured. "I know it would help me keep where I can afford to pay my taxes."

David Allen's amendment was voted down.

Next a motion was made to substitute $650,000 for Drew's figure.

"You must be getting hungry," Coffin commented. "I'm going to call for a division of the house." The vote was close: sixty-six ayes, sixty-nine nays. "The motion to amend has failed," Coffin said.

Finally, a motion was made and seconded for $635,000. Another standing vote was taken, and the ayes took it, eighty-one to forty-nine.

"We now have an amended motion before us," Coffin said, bringing the situation back into focus. "If this is voted yes, that is our budget. If no, that's it for the day." The town would have to get by without a budget until a special town meeting could be called, and that could take a month. Nobody wanted to do that.

"All in favor?" A subdued crowd of ayes.

"All opposed?" A few scattered nays.

"The motion is passed," Coffin declared, and the meeting adjourned for lunch. Everyone filed out the doors and headed for the Methodist Church for soup and sandwiches. They would vote on the rest of the articles in the afternoon.

"What just happened?" I asked Coffin, feeling puzzled. "The final decision on the budget was so subdued."

"Well, I think the group sensed that it had reached an accommodation," Coffin said. "The feeling was, we're not going to give the select board what they want. And the board, for their part, wasn't vocal enough in support of their budget. Their feeling was, if we hunker down, maybe they won't do too much damage."

That's the way it often happens at town meetings. Residents use amendments to feel out which way the group as a whole is leaning, and then they work toward a decision everybody can live with. Through five separate votes, the citizens of Bradford zeroed in on the final amount of their new town budget. The people got what they wanted.

Ruling the Unruly

A week before Bradford's town meeting, Larry Coffin drove thirty miles through a light snow from his house in Bradford to the Elks Lodge in Montpelier. The Vermont League of Cities and Towns was sponsoring an all-day workshop there for moderators from around the state to prepare them for any problems they might encounter at their local meetings. Since Coffin had served as Bradford's moderator for nearly four decades, he knew more about the job than almost anyone in Vermont. But he figured even he could use a tune-up.

Most of the seventy or so men and women who braved the snow to attend the workshop were experienced moderators or local officials, though there also were a handful of first-timers elected by their communities the previous winter for whom the prospect of running a meeting brought a mix of excitement and dread. Being a moderator, after all, is like nothing else in town life. Given sole authority to run the annual meeting, a moderator must not only encourage citizens to participate, but also enforce strict rules. As we saw in Bradford, it's the moderator's job to call for motions on questions, recognize citizens who wish to speak, guide the group through amendments, announce the results of votes, and generally shape the emergence of the town's collective wisdom. Yet a moderator must always remember that the will of the people comes first, which demands some humility.

"I was bragging about being elected moderator," one rookie said, "then one woman took me aside and said, whoever speaks the most but says the least gets elected moderator."

Nobody ever promised it would be easy. Vermonters are by nature an independent bunch who are used to having a say in matters that affect them. Emotions sometimes flare. A few years ago, a moderator named Ed Chase was running a meeting in the village of Westford when a debate got hot. "One crusty old gentleman was jabbing his finger at another citizen and he finally suggested that the two of them 'take it outside and settle their differences right now,'" Chase says. "Well, as moderators we're supposed to follow Robert's Rules of Order, but for the life of me, I couldn't find that anywhere in the book."

The moderator's most important responsibilities, as the group in Montpelier was reminded, are to make sure that every citizen is given the chance to be heard, that deliberation is fair to the group as a whole, and that the meeting makes decisions in a timely fashion. These duties support the basic principles for good decision-making evolved by the bees: *Seek a diversity of knowledge. Encourage a friendly competition of ideas. Use an effective mechanism to narrow your choices.* Fortunately for moderators, they're not expected to rely on their ingenuity alone to achieve these goals. In most cases, they can lean on Robert's Rules, which were written for just these purposes.

What originally prompted Henry Martyn Robert to publish his "pocket manual" for meetings in 1876, when he was a U.S. Army engineer, was his dissatisfaction with the haphazard manner in which meetings—from church groups to civic organizations—were commonly run. His aim was to simplify parliamentary procedures enough to make meetings both effective and fair. To him, that meant protecting the rights of the minority, while enabling the majority to prevail. People should never leave a meeting feeling they haven't been heard, he believed, or that a decision has somehow been railroaded.

Take the first goal, for example, about hearing everyone's opinion. The simplest way to accomplish that, as the workshop instructors pointed out, is to follow Robert's rule that no citizen may speak twice on an issue until everyone else who wants to speak has had a chance to. That makes it difficult, in theory, for a blowhard to dominate debate, or for two individuals to get into a prolonged personal argument. (In practice, of course, things can be looser. Larry Coffin, for one, prefers to let a conversation go without interfering if it seems to be advancing the group's understanding of what's at stake.)

By looking out for the rights of individuals to express themselves, the moderator can also make sure that the group is presented with a broad range of opinions. That's where diversity of knowledge comes in. As the economist Scott Page proved with his "diversity prediction theorem," it can be just as important for a decision-making group to possess a wide variety of outlooks as expert knowledge. In fact, it may be even more important for town meetings, since citizens are often asked to vote on questions that don't have a single correct answer, such as, how much should we spend on the town budget? In such cases, a broad sampling of opinions is essential.

Do town meetings actually promote such dialogue? How many people speak at a typical meeting? More than you might think, says Frank Bryan, a political scientist at the University of Vermont who has attended more town meetings than anyone else alive. Between 1970 and 1998, Bryan and his students collected data from more than fifteen hundred meetings from one end of the state to the other. The average meeting, they found, has 114 people, or about 20 percent of the average town's eligible voters.

Of those who attend, nearly half speak at least once—a surprisingly large participation. "The old-timers tend to speak more than the newcomers," Bryan says, "because they tend to know more. These people get a reputation. When one of them stands up to speak, someone might think, I don't really like the old fart, but he does live in that neighborhood, so he probably knows what he's talking about."

What about the moderator's second goal of promoting a fair and civil competition of ideas? Are moderators able to prevent bullies from dominating conversations? Do they discourage white-collar types from lecturing blue-collar types (or the other way around)? What about men versus women? Here again, Bryan's research is encouraging. There's little evidence, he says, that the small-town dynamics of the traditional floor meeting stifles discussion or debate. "Oh, there are always one or two people that talk too much," Bryan says. "But they don't dominate the meeting or anything. In my hometown we had someone who used to speak too much, but after a while he got the message." The idea that people are reluctant to voice their opinions because they're worried about what their neighbors might think just isn't true, he says. And women speak out almost as frequently as men.

How about the third goal, of effectively narrowing the group's choices? Do town meetings provide citizens with a mechanism to make good decisions? That depends, Bryan says, on how well the moderator follows procedure. "When you read Robert's Rules you say, my God, this is going to be awful," he says. "But when you live through them you realize they're absolutely essential.

Most newcomers are surprised by the degree to which a town meeting sticks to its guns on issues of due process, and bends over backward to ensure that minority rights are observed. Many of them are put off by the tedium of it. They say it with their eyes: Do we have to go through all this? The answer is yes." By following the rules, a moderator can also keep the group from rushing a decision too quickly, which is the source of so many mistakes.

When a meeting is going well, it follows its own inner logic, says Paul Gillies, a veteran moderator from the town of Berlin. "You know how, in a bad discussion, I'm not listening to you and you're not listening to me? I say my thing, and then you say your thing, and then I say my thing again? That doesn't happen in a good town meeting. There's a linear course for the logic of what comes out. Somebody will say, I don't understand what this is about. Somebody will answer it. Somebody will then say, but there's one thing that bothers me about this. And somebody will answer it. You don't get those questions again." Different people, in other words, add different pieces to the puzzle, and there's an upward movement to the discussion. "I always feel as though, by the time we get to the vote, everything that should be said has been said, in a kind of mysteriously correct order, and the group is unified in what it wants to know," Gillies says.

When the talking is finished, and it's time to make a decision, there can be an inner logic to the voting, too. Consider the way the citizens of Bradford used amendments to calibrate their preferences about the town's budget. This was the order, remember, of their proposed amounts:

(1) $628,425
(2) $656,000
(3) $590,634
(4) $684,919
(5) $635,000

The first proposal, for $628,425, came from cost-cutter Larry Drew, who scrutinized every line of the select board's recommended budget and trimmed it by 8 percent. He set the anchor for the rest of the debate. The second proposal, for $656,000, split the difference between Drew's budget and the select board's. The third proposal, for $590,634, was last year's budget, effectively putting a floor on how low the group would go. The fourth proposal, for $684,919, was the amount the select board had requested, putting a ceiling on how high the group would go. The fifth and final proposal, for $635,000, not only split the difference between the highest and lowest amounts, it was also the average of all the proposals. This was the number the voters approved, representing, as it did, the wisdom of the crowd that day.

Just as Tom Seeley's bees, in other words, were able to zero in on the best of the five nest boxes on Appledore Island—with all five choices being promoted by scout bees during several hours of deliberation—so this group of citizens in Bradford worked their way through a broad sampling of options to find an acceptable compromise. Thanks to the structure of their town meeting, which was based on Robert's Rules of Order, and Larry Coffin's skills as moderator, they used their diversity of knowledge about what the community needed to combine their individual preferences into a single number they could all live with. They accomplished what they had come for.

During the past few decades, this kind of face-to-face deliberation has been challenged in Vermont by powerful trends in contemporary culture. As town populations grow and people have less free time, attendance at town meetings has fallen. As a result, some towns have adopted what they call the "Australian ballot" system, in which citizens don't have to come to meetings, but rather fill out preprinted ballots like those used during presidential elections. This system, supporters say, is more convenient, since it takes only a few minutes, and it enables more people to weigh in on questions that affect them.

What's lost in the process, of course, is the give-and-take of deliberation. Voting by preprinted ballot is a yes-or-no proposition. It doesn't provide the community with any more information than that. For certain kinds of votes, such as the election of town officers, that might work just fine. But for others, such as how much a town should invest in its senior center, how much in its library, how much in bike trails, a chance to reason together may be critical to making the right choice. By talking face-to-face, a community gives itself an opportunity to build on its diverse knowledge—like Dennis O'Donoghue's problem-solving groups at Boeing—as different citizens contribute different pieces of the puzzle.

Does that mean there's still a need for town meetings in our increasingly complex world?

"I think that's clear," says Frank Bryan. "The probability of getting a smarter decision goes up dramatically if you hold a town meeting. You get more information at a meeting and, if there are differences, they get exposed. Opinions get tested." Citizens of small towns, in the end, aren't afraid of dealing face-to-face with difficult issues, even if they produce open conflict. "People get pissed off, and sometimes hurt feelings last for years. But life goes on," Bryan says.

That's the simple promise of town meetings. When divisive issues threaten to pull communities apart—whether the issues are intensely local, such as school bus routes or whose road gets plowed first, or national in scope, such as the war in Iraq or the battle over gay rights—town meetings offer a process to bring people back together. Citizens who talk to one another give themselves a better chance to make smart decisions.

Honeybees and Same-Sex Weddings

A few minutes after midnight on July 1, 2000, in the town of Brattleboro, Vermont, Annette Cappy, the town clerk, handed to Carolyn Conrad and Kathleen Peterson a piece of paper that made history. The document was a license for a civil union between two people of the same sex—the first in the nation. A few minutes later, the couple exchanged vows before a justice of the peace, and a small crowd cheered. From that moment on, Conrad and Peterson would share the same rights and responsibilities as married people, such as taking care of each other financially, sharing property, or making medical decisions for each other. Whenever they filled out forms that asked for spouse, immediate family, dependent, or next of kin, they'd know that meant them, too. They were no different now than anyone else in the eyes of the law.

Not everybody in Vermont was happy about that.

"Some people were convinced it would be the end of marriage as we know it," says Deborah Markowitz, Vermont's secretary of state. The public debate grew so intense, in fact, that it threatened to tear some communities apart. Town meetings broke out into arguments. A few farmers painted angry slogans on their barns. "Some hateful things were said."

It began in December 1999, when the state supreme court ruled that Vermont laws that prohibited same-sex marriages violated the state constitution. The court ordered the state legislature to correct the situation, but the lawmakers had already gone home on recess, which set the stage for fierce local debates. "The voters didn't really care that the legislature was being told it had to act by the courts," Markowitz says. "People were saying, ignore the courts. Impeach the judges. Whatever. It was a hard beginning to the debate."

At town meetings in March, residents vented their feelings. "Some things were difficult to hear, like, we don't want those people in our community. We don't want to know about you. You're going to affect our community in some way," Markowitz says. "There was a movement called Take Back Vermont, where people put out big signs urging old-timers to take Vermont back from the flatlanders who'd come here since 1960, all those liberal newcomers."

But something else was happening, too. The debate was starting to shift as people realized that members of their own communities were living in gay families. "People started standing up at town meetings saying, you know, I've been your neighbor forever and I'm gay," Markowitz says. "Then other people started thinking, 'Oh, you don't actually have horns,' because it's hard to stereotype somebody you've engaged in civic discourse with." Eventually, the hard-liners' position shifted from "Over my dead body" to "I don't really want to know about it," which was enough for life to move on. A few weeks later, the legislature passed the civil union bill, and Governor Howard Dean signed it into law.

Since then, more than seventeen thousand people have been joined in civil unions in Vermont, and the controversy has more or less gone away. In fact, beginning in September 2009, same-sex couples have been able to enter into marriages in Vermont, eliminating the need for civil unions. "When surveys ask Vermonters about same-sex couples now, a vast majority say it's a nothing burger," Markowitz says. "Some people even admit they feel foolish about the way they acted."

How did this happen? How did people in Vermont get over their deep differences? The answer, according to Markowitz, has to do with what Vermonters take away from town meetings. "The discourse that happens in a town meeting, even though there may be winners and losers, even though there might be hard words spoken, can enrich people's ability to help each other in times of trouble," she explained. That's why settlers went to town meetings in the first place. "Back then they went not only to decide how much firewood everyone was going to give to support the local school or how many hours they were going to give to help build roads, but also to meet their neighbors so that if their barn burned down and they had to raise a roof, they would know somebody, because they lived in very isolated homesteads."

In a way, Vermonters still do, she says. "We still live isolated lives. Not isolated by distance or lack of transportation, but by modern media and culture. Everyone spends time in their houses in front of their televisions or commuting to work." But there's still also value in community, she says, and the fact that we rely on each other.

A few years ago, Robert Putnam, a political scientist at Harvard, discovered what he called a "whimsical yet discomfiting" piece of evidence about how disengaged Americans have become from civic affairs. He was talking about everyday activities like PTA meetings, church groups, or Lions Club get-togethers that build up a kind of "social capital" that a community can count on during times of crisis. What Putnam learned was that the number of Americans who bowled had gone up by 10 percent in recent years, while league bowling had decreased by 40 percent. More Americans, in other words, were bowling alone. "The broader social significance of this," he explained, "lies in the social interaction and even occasionally civic conversations over beer and pizza that solo bowlers forgo." As bowling leagues withered, so did communities.

The fact that a growing number of Americans had turned their backs on civic life was a sign that something had been lost, Putnam says. Something important had been squandered. People, unlike honeybees, don't possess a natural instinct to serve the collective, but, as intensely social animals, we do understand the power of belonging, and we miss it when it's gone. That's what Vermont town meetings had offered during a painful debate: the kind of empathy for neighbors and fellow citizens—whether you agreed with them or not—that gave back a portion of what had been lost.

"It's important not to idealize town meetings, because they don't always work perfectly. A lot of factors have to be in place for a meeting to enrich civic life in a community," Deborah Markowitz says. "But because of our ability to engage in civic discourse, the divisions over civil union have healed and mended quickly. Remarkably quickly. Not everywhere. And not with everyone. But the conflict and strong feelings didn't endure. The town survived. The people survived."

The hive adapted and moved on.

3

Termites

One Thing Leads to Another

The first power line to fail ran between the Harding substation south of Cleveland, Ohio, and the Chamberlin substation twelve miles away. At 3:06 p.m. on August 14, 2003, the line brushed up against the branches of an overgrown locust tree and shorted out. For the local power company, FirstEnergy Corp., which owns thousands of miles of transmission lines, this wasn't unusual, especially in late summer. On a normal day, it wouldn't have caused much of a fuss. Power would have been rerouted, a crew would have been sent out to check the damage, and life would have gone on.

But this wasn't a normal day.

Earlier that afternoon, a coal-fired generating unit at the Eastlake power station on Lake Erie had tripped, taking a portion of FirstEnergy's reserve off-line. With two other power plants shut down for maintenance, and air conditioners throughout northern Ohio trying to keep up with a high of 87 degrees, that put FirstEnergy's system into an uncomfortably vulnerable position, relying more than usual upon power from outside the region. That, in turn, put a greater strain on the transmission lines carrying the power.

What happened next was a vivid demonstration of how a complex system behaves. With thousands of generators and millions of miles of transmission wires, the power grid in North America is one of the most elaborate networks ever created—far more interconnected, in its way, than the largest colony of ants or bees. When something happens in one part of the grid, no matter how insignificant it may seem, the impact can be felt in places far, far away.

At 3:32 p.m., a half-hour after the Harding–Chamberlin line failed, another high-voltage line failed, this one between the Hanna substation south of Cleveland and the Juniper substation east of Akron. The reason, again, was contact with a tree. By chance, a tree-trimming crew was in the neighborhood and witnessed the event. Nine minutes later, a third line tripped, between substations in Akron and Canton, also from a tree. This wasn't coincidental, because the line had been overloaded by power diverted from the earlier failures, causing it to heat up and sag. Now the region's system was teetering on the edge.

The final straw came at 4:06 p.m., when a high-voltage line tripped between the Star substation near Akron and the W. H. Sammis power plant on the Ohio River to the southeast. Unlike the previous incidents, this one wasn't caused by contact with a tree, but by unsafe fluctuations in voltage created, in part, by the earlier transmission line problems. Protective devices on the line called impedance relays sensed a power surge coming through the system and tripped circuit breakers to keep the line from harm. Unfortunately, the Sammis–Star line happened to be the last link connecting the Cleveland region with the rest of the power grid to the east. When it went down, it started a chain reaction of failures that spread north and west, tripping power lines in northern Ohio and southern Michigan and blacking out Detroit. Then the cascade suddenly reversed direction, following a powerful surge across Pennsylvania into New York and Ontario that raced counterclockwise around Lake Erie, knocking out transmission lines and power plants. Within six minutes, more than five hundred power plants in eight states and two Canadian provinces went down, cutting off electricity to some 50 million people.

It was the worst blackout in North American history.

The billboards in Times Square went dark at 4:11 that afternoon. At the same moment, traffic signals failed, subway trains halted, restaurant freezers quit, sewage treatment plants stopped working, and hotel guests got locked out of their rooms all across Manhattan. Coming less than two years after the attack on the World Trade Center, the blackout temporarily stunned New Yorkers. But as soon as they learned that the emergency had nothing to do with terrorists, they set out to make the best of a difficult situation.

At least 350,000 people, at that point, were stranded on trains or subways, some on bridges or deep inside tunnels. Passengers on a train trapped under the East River were rescued when a diesel locomotive pulled them back to Penn Station. Thousands of others were forced to fend for themselves, walking single-file along the darkened tracks to the nearest station.

With city streets in gridlock, throngs of commuters trooped home across the Brooklyn Bridge. Cell phone systems had gone down almost immediately, swamped by callers trying to reach family and friends. Water pressure disappeared in high-rise apartment buildings, prompting residents to line up at fire hydrants for clean water. Automatic teller machines refused to give out cash. At Mount Sinai Hospital on the Upper East Side, a patient missed her chance for a liver transplant when surgeons couldn't guarantee conditions in the operating room. "It wasn't safe to start a transplant if we didn't know whether we could finish it," one of the doctors told reporters.

Elsewhere citizens were pitching in. Three blocks from the Empire State Building, at the intersection of Thirty-fourth Street and Lexington Avenue, a sixty-year-old salesman in a dress shirt and silk tie was directing traffic for the first time in his life. A driver passing by handed him an empty plastic bottle to use as a baton. "You'll need this," she said.

The city had suffered big blackouts before in 1965 and 1977, and preventive measures had been taken after each one. "It wasn't supposed to happen again," Governor George Pataki said at a press conference, "and it has happened again. And there have to be some tough questions asked as to why."

Nobody at the time, of course, had a clue what had taken place. In Canada, a spokesman from the prime minister's office pointed a finger at New York, saying the blackout had started when lightning had hit a power plant near Niagara Falls. In New York, an official pointed his finger at Ohio, saying there'd been a problem at a nuclear power plant on Lake Erie. It took months for a federal task force to get to the bottom of it all, and when investigators did, they discovered a few surprises.

It wasn't just the tree strikes that had led to the blackout, after all. A computer glitch in the FirstEnergy control room near Akron had also played a role, turning off the automated alarm system at 2:14 p.m. on the day of the blackout. That meant that operators in the control room didn't notice when the first transmission line shorted out less than an hour later. After the second line went down at 3:32, someone in the regional monitoring center called to ask about the failure, but a technician in the control room was surprised to hear about it.

"Doggone it. When did that happen?" he asked.

Meanwhile, other calls were pouring in from industrial customers, neighboring utilities, and FirstEnergy's own power plant operators, who were all trying to interpret signs of trouble on the grid.

"I'm still getting a lot of voltage spikes and swings on the generator," said an operator at the Perry nuclear power plant in Ohio, who was worried about his unit shutting down automatically. "I don't know how much longer we're going to survive." Only after the lights went out in the FirstEnergy control room did operators know for sure that it was their system and not somebody else's that was about to collapse. By then it was too late.

How could this have happened? How could a tree touching a

power line near Cleveland and a computer glitch near Akron have triggered an event that caused more than $6 billion worth of damage? The cause and effect didn't seem to match up. If terrorists had set out to accomplish such a task, as New Yorkers had initially feared, they couldn't have done so devastating a job. How could such a small incident have created such a colossal mess?

"This blackout was largely preventable," says Spencer Abraham, then secretary of energy, when the task force presented its findings. "A number of relatively small problems combined to create a very big one."

But that wasn't quite right.

It wasn't just bad luck or poor maintenance that had caused an unlikely series of glitches to take place. It was the structure of the grid itself that was to blame. Cobbled together over more than a century, the grid had become so complex, experts said, its billions of parts so interconnected, that sooner or later a massive failure like this was certain to happen. In a sense, "the entire electrical grid in North America was just one big electric circuit," wrote Thomas Overbye, an electrical engineer at the University of Illinois at Urbana-Champaign. "The humble wall outlet is actually a gateway to one of the largest and most complex objects ever built." The upside of this connectivity is that, during periods of unusual demand, utility companies can share power with one another, which keeps costs down and makes sure the lights come on when we flip the switch. The downside is that failures in one part of the system can spread rapidly to others. A highly connected system can fail, Overbye wrote, "and when it does it fails in complex and dramatic ways."

The 2003 blackout was a prime example of what network scientists call a cascade, in which an initial event—like a spark that starts a forest fire or a financial report that triggers a stock market crash—makes other events more likely, which make still other events more likely, and so on, until a cascade spreads through the whole system. One thing, in short, leads to another. "The trouble with systems like the power grid," writes network scientist Duncan Watts in his book, *Six Degrees: The Science of a Connected Age*, "is that they are built up of many components whose individual behavior is reasonably well understood (the physics of power generation is nineteenth-century stuff) but whose collective behavior, like that of football crowds and stock market investors, can be sometimes orderly and sometimes chaotic, confusing, and even destructive."

What does this have to do with ants, bees, or other insects? What could such tiny creatures possibly have to teach us about keeping the power grid up and running? The answer is that natural systems like smart swarms have evolved specific behaviors to avoid such cascades, the Achilles' heel of highly connected networks. In an ant colony or a beehive, many individuals can fail to perform their jobs and the system still functions just fine, because many other individuals, sensing something different in their surroundings, adjust their behavior accordingly. In a way, such a system is self-healing.

A few years ago, the Electric Power Research Institute in Palo Alto, California; the Department of Defense; and several other organizations set out to determine if the North American power grid could be made self-healing, too. The first step, researchers concluded, would be to make each component smarter. "Every node in the power grid should be awake, responsive, and in communication with every other node," wrote Massoud Amin and Philip Schewe in *Scientific American*. To give the grid the same capability that ants and bees possess for instantaneous damage control, that is, every component in the system—every breaker, switch, transformer, bus bar, and transmission line—would need its own processor to monitor conditions and optimize performance, without having to first check with a human controller. The grid would need to develop a distributed intelligence.

The second step, they wrote, would be to give the grid a rapid means of predicting different kinds of problems. To do that, the researcher recommended giving the system the same kind of "look ahead" capability that we saw Arthur Samuel create for his checkers-playing program in Chapter 1. Just like a master checkers player, that is, who plans several moves ahead, the grid needs to be able to develop what-if scenarios for the near future. The third and final step would be to split the grid into "islands" to isolate failures in the event of an emergency, thus preventing them from spreading across the whole system—a strategy we'll see later in this chapter that termites use when their colony's nest mound is attacked.

Would a self-healing grid have prevented the kind of massive blackout we saw in 2003? Maybe so, Amin and Schewe speculate. If sensors in sagging transmission lines had detected an abnormal power flow, they might have redirected power to other lines hours before they failed. At the same time, look-ahead simulators might have predicted the probability of these failures and recommended corrective action to human operators. Instead of a catastrophic loss of power, they wrote, "the most a customer in the wider area would have seen would have been a brief flicker of the lights. Many would not have been aware of the problem at all."

Which is exactly what we want, after all, because we've come to rely so completely on complex systems like the power grid—or our transportation networks, or the stock market, or the Internet—to support our way of life. By enabling us to share essential resources and information with one another, such structures make our society possible—just as structures built by social insects, from pheromone trails to termite nests, make their societies possible. Unlike our systems, though, which tend to crash from time to time, those created by smart swarms tend to be less brittle and more resilient.

One reason why can be found in a dusty pasture in southern Africa.

A Castle Fit for Millions

Standing ten feet tall or higher on the savannah, they look like witches' hats made of dirt. Because of the way they're tilted, you might wonder who left these conical towers of mud out here and what they mean. Local folks call them "ant heaps," but they don't have anything to do with ants. These pointy structures were erected by termites, and, despite their odd appearance, they represent one of nature's most sophisticated architectural achievements.

For the past few years, J. Scott Turner, a biologist from the State University of New York in Syracuse, has been coming to the Omatjenne Agricultural Research Station in northern Namibia to study the complex structure of these mounds. They're all over the station, one or more per acre, evenly spaced across the sixty-five square miles of acacia *thornveld* where researchers graze goats, sheep, and cattle as breeding stock for Namibia's commercial and communal ranchers. "There's definitely no shortage of mounds to work on," Turner says. "This is termite central."

Naturalists have known since the eighteenth century what's inside these mounds. "The exterior is one large shell in the manner of a dome, large and strong enough to inclose and shelter the interior from the vicissitudes of the weather, and the inhabitants from the attacks of natural or accidental enemies," reported English explorer Henry Smeathman in 1781, describing the mounds as if they were castles. The interior, meanwhile, contained numerous apartments for the king and queen, nurseries for their progeny, and storerooms filled with provisions, all of which, he wrote, are "contrived and finished with such art and ingenuity, that we are at a loss to say, whether they are most to be admired on that account, or for their enormous magnitude and solidity." If such mounds had been built at a scale for people instead of insects, he noted, they would be nearly five times taller than the Great Pyramid.

Considering the monumental size of the mounds, it's surprising that so small a part is taken up by a colony's nest. In a typical mound built by *Macrotermes michaelseni*, the species Turner studies, the colony's roughly spherical nest, which might measure three to four feet in diameter, is normally located just below ground level. That's where the queen's chamber can be found, surrounded by nurseries for the young and extensive galleries for growing fungus. Like leaf-cutter ants, these African termites cultivate a particular species of symbiotic fungus (*Termitomyces*) to help them break down raw wood and grass into something they can digest. The colony needs a great deal of this fungus to feed the two million or so termites that live inside a large nest. Everything else aboveground, including the lofty tower, performs an entirely different function: to regulate the atmosphere inside.

Unlike the harvester ants we met in Chapter 1, which are covered with a greasy coating that protects them from desert heat, termites have a tender skin and must avoid getting dried out. Even in the arid climate of Namibia, they need a humid environment to be healthy—a relative humidity of around 90 percent, to be specific. Also, the termites and their fungal compost piles must breathe, just as we do, and at the same rate that people do, perhaps more. Yet they must breathe while living underground in the nest. Just as would be the case if we were buried underground, things could get very stuffy very quickly if something were not done. And the mound is what the termites do, building it as a device to capture wind to ventilate the colony, conveying carbon dioxide and water to the mound surface, where it is lost through a network of narrow channels connecting ultimately to the world through tiny "egress tunnels." There, oxygen is picked up and conveyed to the nest for the termites and fungi to breathe.

The whole termite mound, in other words, isn't just a shelter for the insects. It's more like a giant lung, consuming as much oxygen as a goat or a small cow. Turner considers the mound's function as a respiratory system so essential that the termites couldn't live without it. In a sense, he argues, the mound is almost a living part of the colony.

To better understand the way a mound works, Turner and Rupert Soar, an engineer then at Loughborough University in the United Kingdom, set out a few years ago to make a digital 3-D model of one. The first stage of their project involved creating a plaster cast of a mound's interior, which meant coating the exterior with a hard shell of plaster, then filling the passageways inside. This wasn't the first time such a feat had been attempted (a Belgian entomologist named Jean Ruelle did something similar in the early 1960s), but it was an enormous task all the same. After scaffolding was erected around the mound, a concrete-type mixer was trucked in, along with six tons of plaster. Then, with the help of graduate students and local laborers, Turner and Soar poured bucket after bucket of a liquidy plaster-and-water mix into the mound. This took several days, after which they left the mound alone for six months to let the plaster harden.

The second stage was just as challenging. To convert the mound into a digital model, Soar and his colleagues designed, built, and shipped a special milling and scanning device to Namibia. A metal frame was erected over the mound to hold the machine's slicer. The frame alone stood two stories high. The slicer, which had a circular blade not unlike those used to cut meat in a delicatessen, was hung horizontally on rails over the termite mound. Then the team began slicing away the top of the mound one millimeter (.04 inch) at a time. After each slice, Soar and his colleagues took a digital photograph of the mound's exposed flat surface, showing not only the dark soil but also the white plaster in ventilation passageways. These photographs were the equivalent of the hundreds of digital scans routinely made by medical technicians during computer tomography (CT), which are assembled by software into 3-D images. It was a dirty, tedious process; each slice through the mound's dirt and plaster took ten minutes to complete. Working day and night for two months, Soar and his team collected 2,500 slices of the mound from top to bottom. "It nearly killed us all," he says.

But it was worth it in the end. When they looked at the 3-D image a few weeks later, after the first number-crunching was done on the data, they were amazed by what it showed. "It wasn't until we saw the whole thing that we really understood the magnitude to which the termites are engineering their environment," he says. "It took our breath away." The network of tunnels and air passageways inside the mound was intricate and dense, as complex as a Chinese puzzle ball carved from ivory. Filigreed channels in the base rose toward a point halfway up the mound, where a kind of spire began. Within the spire, two or three large vertical channels rose all the way to the top. As impressive as the structures were, though, what really surprised Turner and Soar was that the structure showed how termites use wind energy in an entirely new way.

Natural winds are inherently messy, changing speed, location and direction all the time—in a word, they're turbulent. For the most part, engineers and architects regard this messiness as a nuisance, trying to minimize it or escape it altogether, and this is why wind power currently comes mostly in two flavors: either using large turbines to turn wind into electricity, or designing tall buildings that take advantages of gradients, or differences, in wind speed from the ground up. Both are expensive propositions. But termites, Turner and Soar realized, use their mounds to turn the "nuisance" energy in turbulent winds into wind-power gold. The structure of the mounds was crucial to this ability. The large tunnels inside the mound formed a network of organ pipe–like chambers that resonate when turbulent winds blow past the mound. The mound doesn't sing, as an organ might, so much as it hisses, ranging from a comparatively high frequency sibilance at the surface to a low frequency sigh at the center of the mound and nest. This helps mix the stale nest air with the wind-refreshed mound air, ultimately enabling the nest to breathe.

Soar was so impressed by termite mounds as adaptable "smart structures," he wondered how to apply their basic principles to create energy-efficient buildings for people. "What's interesting to me about these structures is that they achieve essentially the same level of comfort for the termites as we would expect in our homes, but they're not using electricity," he says. "The termites use wind. They use solar energy. Maybe we can adapt some of their self-regulating or energy-controlling abilities for our own buildings."

One way to do that, he believes, would be to imprint organic-like designs into walls to change them from being a solid piece of masonry to being "permeated by something akin to channels that you'd find in blood vessels." With new methods for printing complex three-dimensional structures, it's now possible to replicate in buildings the termites' functional tricks with turbulent winds. Such structures could be embedded directly into the walls of a house, he says. "So instead of being solid brick or concrete, we're talking about shapes and contours to act like membranes and achieve the same regulation by taking advantage of energy gradients—without using electricity." Instead of being an impermeable barrier between outside and inside, the wall could become a device for managing the flows of matter and energy that control the climates inside our buildings—all through the clever use of wind, which Turner and Soar hadn't realized was possible until the termites showed them how.

As IMPRESSIVE as termite mounds are from an engineering point of view, they're truly baffling from a biological perspective. How can creatures no more than a quarter-inch in length, with minimal brainpower and little or no vision create structures of such complexity and sophistication? How do they know how tall a tower should be? How do they know where to place each of the mound's narrow conduits? Unlike human architects, termites have no blueprints to follow as they put together their castles of mud. How do they know if their lungs of clay will get rid of enough carbon dioxide or retain the right amount of moisture?

It normally takes a new colony between four and five years to build a mound. During that time, several generations of workers will be born and die. But when Turner and his crew, as an experiment, knocked down one mature mound with a front-end loader, the colony replaced it within ninety days. The restored structure, the researchers confirmed, was just as efficient as the old one in regulating gas flows and temperature. How did the termites know what was missing? They couldn't have remembered the way the old mound looked. What secret ability do termites have that enables a colony to build something as a group that none of the insects could possibly imagine alone?

Part of the answer to this mystery was uncovered during the 1950s by Pierre-Paul Grassé, a French biologist who wasn't even thinking about termites when he first visited Africa. A former medical doctor, Grassé had been studying parasites that live in the guts of termites. But he soon became fascinated with the insects themselves and the remarkable way that they collaborate to build things.

Grassé noticed that individual termites are quite sensitive to changes in their environment. If a worker carrying a grain of soil, for example, comes across a small pile of dirt left by fellow workers, it will drop its grain on the pile. That action, in turn, stimulates other workers to do the same, and pretty soon, if there are enough termites around, the small pile of dirt grows into sizable pillars. Grassé called this process *stigmergy*, which he defined as "the stimulation of workers by the performance they have achieved."

Instead of interacting directly with other individuals, in other words, termite workers interact with the structure they're building together. As that structure grows and changes, so does the manner in which they interact with it. The structure itself becomes their guide. Once a pillar reaches a certain height, for example, workers stop making it taller and start building an arch sideways to connect it to other nearby pillars. That, in turn, becomes the foundation for a new wall inside a termite mound.

This process of stigmergy, as Grassé described it, is yet another form of self-organization. By following a simple rule of thumb—"Drop your grain of soil here if somebody else has already done so"—termite workers are able to build something together that none of them can comprehend alone. Just as foraging ants are able to find the shortest path to a food source and scout bees can choose the best site for a new home, so termite workers have evolved a way to create a complex shelter for the colony. The key difference is that when termite workers interact with one another during the construction process, they do it *indirectly* rather than face-to-face.

This seemingly small difference has a big impact on the efficiency of the system, says chemist Jean-Louis Deneubourg, whose research on ants we looked at in Chapter 1. "Imagine that two termites must meet each other to decide to drop a pellet," he says. "If I must wait for you to come and say, oh, drop it here, that could be very costly. You might have to walk twenty centimeters [eight inches] away to collect another. I could be doing something else. But if we can interact through the byproduct of our work, then I don't care what you're doing. I don't care where you are. Our interaction is not direct, but indirect."

What we're talking about here is a surprisingly simple mechanism to enable *indirect collaboration* on a massive scale. If individuals in a group are prompted to make small changes to a shared structure that inspires others to improve it even further, the structure becomes an active player in the creative process.

This can lead to all kinds of new possibilities in the way that groups share information and solve problems. That's why *indirect collaboration* is the third principle of a smart swarm, along with *self-organization* and *diversity of information*.

Just as utility companies over the years pieced together the complex network we call the power grid to distribute electricity, so termites piece together the complex structure we call the mound to distribute gases and moisture. Just as private individuals and businesses have cobbled together the digital network we call the World Wide Web, adding each site to a shared structure that grows more immense every day, so termite workers cobble together the walls and corridors of their insect castles, each one adding to a shared structure beyond their understanding. But unlike our systems, which are tuned for efficiency, the termites' systems have been tuned for robustness, which they demonstrate by building mounds that are constantly self-healing.

To better understand how this works, Scott Turner and his colleagues conducted a series of experiments, in which they intentionally damaged the outer wall of a termite mound. They wanted to see how the colony would respond to a breach of the mound's surface—something that happens fairly often in Namibia during the rainy season, from January to May, when downpours can wash away sections of a mound's exterior. In one experiment, Turner's team used a soil augur to drill a four-inch-wide hole into the side of an eight-foot-high mound. They drilled the hole about a foot into the mound, making sure that it penetrated at least one of the conduits that circulate air throughout the structure. They didn't have to wait long for the colony's rapid-response squad to show up. Within five minutes, oversize termite soldiers appeared along the edge of the breach to stand guard as dozens of workers followed, busily filling up the hole with soil.

The work was rapid and systematic. Each worker carried a grain of soil to the site to add it to the work in progress. Each of these grains was coated with a kind of salivary glue, and this glue contained a pheromone that stimulated other workers to leave their grains at the same spot. As each worker placed its grain on top of other grains, the insect would often rock its head back and forth to make the fit snug, not unlike what a brick mason might do when laying a wall. In this way, the breach was soon filled with what Turner describes as a "spongy" pattern of pillars, walls, and tunnels that served as a temporary patch until more workers could be recruited to finish the job. The whole thing took less than an hour.

Two days later, Turner's team came back with a chain saw and cut off the top of the mound to examine the larger network of tunnels inside. They discovered that the workers repairing the surface hadn't been the only ones launched into action by the emergency. While workers were plugging up the breach, others deeper in the nest had been sealing off secondary passageways leading to the hole, like crewmen on a submarine sealing off a leak in the hull. To Turner, this suggested that something bigger than a local repair job was going on. "There was a lot of stuff happening all over the mound, not just at the site of injury," he says. "And it wasn't just the standard stigmergic type of behavior, but a rich suite of interactions between structures and termites."

The termites, remember, don't inhabit the mound's tower. They live in the nest underground. So what kind of alarm bells had gone off to alert the colony? How did they know the precise location of the breach up in the tower? As Rupert Soar's plaster cast had shown so dramatically, the mound was a complicated maze of passageways. What clues had pointed them in the right direction? For Turner, the answer lay in a better understanding of how the insects responded to *any* changes in their environment.

"We'd done all kinds of experiments and had ended up with results that didn't seem to make any sense," Turner says. "For the most part, I was completely confused by the termites' behavior. What really made the lights go off was the realization that termites don't pay attention to the environment itself, but to *changes* in the environment." As long as conditions remained the same, termites functioned well in a relatively wide range of environments. It didn't seem to matter to them if the air inside their nest had a concentration of 1 percent carbon dioxide or 5 percent carbon dioxide. But if you changed it from 1 to 5 percent, they reacted right away. "That made a whole lot of things fall into place," he says.

When the mound wall was breached, it opened up a portal for disturbance to enter inside. Under normal circumstances, life in the nest was steady and still, because the mound filtered out the impacts of turbulent winds outside. But the hole sent unpleasant waves of disturbance—in the form of changes in carbon dioxide concentrations, or humidity, or air movement, or some other factor—that spread quickly through the mound. "As far as we can tell, that's how termites know something is wrong," Turner says. "They start feeling these fluctuations, and that's what wakes them up. That's when they start moving into the mound."

As the termites scattered through the mound, some encountered more disruption than others, depending on how close they ventured to the damage site. Some may have rushed to the site because they were attracted by the intensity of the disturbance, while others ran into what Turner called "sudden transients" along the way. These termites were probably prompted by these events to lay down a grain of soil right then and there, which would have attracted other termites, and so on. Such positive feedback was strongest, of course, at the site of the breach, where recruitment snowballed as pheromone levels shot up. But even deeper within the mound, wherever termites felt these waves of disturbance from the breach, workers were stimulated to start plugging up passageways.

This is precisely the kind of self-healing process that engineers at EPRI had in mind for the North American power grid. Within moments of sensing a disturbance, responding in a completely distributed fashion, termites throughout the mound sprang into action to isolate and manage the disruption. To get a better feeling for how this works, picture what might happen in an air-conditioned office building in which a window was broken, as Turner suggests in his book *The Extended Organism*.

> One way to deal with a broken window in an air-conditioned building is to isolate the break somehow—close the door to

> the office containing the broken window, for example. Most likely, the person to do this would be someone sitting close to the broken window, since that person would feel the perturbation most strongly. If you add to this scenario an interaction between office workers, such that the sight of one person getting up to close the door to the office with the broken window elicits other office workers to begin closing off other nearby doors, or the halls, or the stairwell doors, then you have something akin to the response of workers termites to a breach of their mound wall.

All this activity makes a racket, as Soar and a sound engineer named Jon Crawley discovered. When they inserted a microphone into a similar mound that had been damaged, they heard a furious tapping inside. "It sounded like this incredibly large dinner party," Turner told a radio interviewer. There was all this clacking and banging as termites hit their heads against the walls, signaling others to bring their soil here, or put their soil there. "Imagine a banquet hall filled with a million and a half people all banging their knives and forks against their plates and you'll get some idea of what it sounds like when they're building this thing," he says.

Before long, as workers patched up the breach and waves of disruption faded away, work everywhere slowed down. In places where the building activity had been less frantic, such as the secondary passageways leading to the damage, jobs were left unfinished. Many would be unplugged and restored to their original function later, sometimes weeks or months later, as the colony resumed its perpetual reshaping of the mound both inside and out.

This was the other part of Turner's revised understanding of how the mound worked. Not only does this complicated structure represent an *indirect collaboration* among millions of individuals, it also embodies a kind of ongoing *conversation* between the colony and the world outside. "The mound might look like a structure, but it's better thought of as a process," Turner says. Its shape might seem unchanging from week to week, but a time-lapse movie would reveal it to be "a churning mass of soil." You might not see evidence of it every day, but erosion strips away about 500 pounds of soil in a year (although in some instances, it can be as much as a ton). That soil is continually being replaced by the termites in ways that meet the colony's changing needs. "Let's say a nearby tree falls down and it gets windier on the mound—the turnover of soil in the mound can be enlisted to change its structure in response to those factors," Turner says. By contrast, if the population of a colony increases and the nest becomes too stuffy, the termites might extend the mound's tower higher to capture stronger winds to help ventilate the nest.

Instead of picturing the mound as a barrier that protects the colony from the world outside, Turner says, we should think of it as a dynamic system that balances forces both inside and outside its walls to create the right environment for the termites. This balancing act, he argues, is a form of biological "homeostasis" not unlike that seen in other physiological systems that regulate critical processes, such as the way a warm-blooded animal maintains a steady body temperature despite the fact that it lives in a cold environment.

That's what sets these mound-building termites apart from other termite species, Turner says. By creating a smart, flexible structure in which to live, Macrotermes colonies make it possible for large numbers of individuals to collaborate effectively. This was the same lesson analysts at the Central Intelligence Agency (CIA) and other government bureaus discovered a few years ago when they realized how much information was being lost every day because of inefficient communication. What they urgently needed to keep up with their world, they say, was a smart flexible structure with which to collaborate more effectively—like a virtual termite mound of information.

A Better Way to Spy

The first reports were alarming. A small plane had crashed into a high-rise in upper Manhattan. It was Wednesday afternoon, October 11, 2006, and memories of the terrorist attack five years earlier resurfaced quickly. Within minutes of the 911 call, fighter jets were scrambling over New York, Washington, Detroit, Los Angeles, and Seattle, sent by the North American Aerospace Defense Command (NORAD). Police with heavy weapons flooded the blocks around the smoking building at 524 East Seventy-second Street to seal off the neighborhood, in case the crash turned out to be something other than an accident.

Shortly after 9/11, officials in New York, unlike those in Washington, D.C., had relaxed aviation rules to allow small planes and helicopters to buzz around near the city. So the fact that a plane had penetrated Manhattan's airspace wasn't proof in itself that the pilot had intended to do any harm. Only after authorities had arrived at the scene, talked to eyewitnesses, and sorted through the wreckage did the surprising truth emerge.

The aircraft, a small single-engine private plane, had struck the Belaire, a forty-two-story condominium apartment building overlooking the East River. The engine, propeller, and nose landing-gear strut had lodged in an apartment on the thirty-second floor, where they'd ignited a fire. The rest of the plane had tumbled to the street, carrying with it the bodies of two men, who were soon identified. A passport found at the scene belonged to Cory Lidle, a well-known pitcher for the New York Yankees. The other man was Tyler Stanger, his flight instructor.

Bit by bit, a picture of the accident came into focus at the scene. Construction workers on the top floor of the Belaire had seen the plane angling toward them. "It was coming right at us," one of them told a reporter for *The New York Times*, saying the plane had been so close he could see the pilot's face. "The whole building shook. Then we ran for the elevator." A woman on the thirtieth floor was knocked to the floor by the impact, which blasted out her apartment window and seared her back with flames. She was taken to a nearby hospital to be treated. Down below, a woman on a stationary bike in a gym facing the street saw the wreckage of the plane coming down. "Pieces of it were falling to the sidewalk," she says. "It was aluminum and it was smoking."

Meanwhile, throughout the virtual halls of the United States intelligence community, another kind of fact-gathering activity was taking place. For six months, analysts at the nation's sixteen intelligence agencies had been experimenting with a new network tool called Intellipedia. Based on the same collaborative software as the popular online encyclopedia Wikipedia, the new system gave analysts a way to post pages about subjects of interest on a shared platform on the federal government's classified network, where others in the intelligence community could then add bits of data, correct mistakes, or refocus questions in any way necessary. As a platform for *indirect collaboration*, it was a spy's version of a smart adaptive structure.

Within twenty minutes of the plane crash in New York, agents from across the intelligence community had created a page about the event on Intellipedia. More soon joined in, describing what was known so far. "Within two hours of the crash, Intellipedia's crash page was updated eighty times," one analyst told an agency newsletter. This wasn't an official investigation, by any means. No one ordered the analysts to get to the bottom of the situation in New York. Most of the information was coming in from "open sources" available to the public, such as radio broadcasts, Web videos, and Internet discussion boards. It soon became apparent, though, as officials on the scene were also concluding, that the crash was nothing more than an accident. What was different about the process on Intellipedia was that it was being done in the open through collaborative software. For some old-time analysts, that required a new way of thinking about how intelligence should be gathered and analyzed.

The idea for Intellipedia is generally credited to a CIA analyst named D. Calvin Andrus, who argued in a 2004 paper that American intelligence agencies needed to become more flexible, adaptive, and self-organizing, like complex systems in nature. Just as individual ants in a colony decide which task to perform, he wrote, "so, too, intelligence officers must be allowed to react—in independent, self-organized ways—to developments in the National Security environment." Citing the work of biologist Deborah Gordon, whom we met in Chapter 1, Andrus noted that "adherence to simple rules at an individual level allows ant colonies at the group level to respond to both strategic (seasonal) and tactical (predatory) changes in their environment." Similarly, intelligence agencies needed to keep up with changes in their world, he wrote, through "self-organizing knowledge websites, known as wikis, and information-sharing websites known as blogs."

When Andrus met with a test pilot group of analysts at the CIA to explain his vision, he raised a lot of eyebrows. "We all looked at him and said, this guy is crazy. This guy is certifiable," says Sean Dennehy, a CIA analyst, speaking at a technology conference in Boston with his colleague Don Burke. "We just went through 9/11. We just went through Iraq WMD. And now this guy says we can go and make edits in the middle of the night on a wiki page and change exactly what it says?"

As soon as Dennehy took a closer look at how Wikipedia

actually works, however, he began to see its appeal. Every entry in Wikipedia has an accompanying page for "discussion," where additions and corrections to the entry are vigorously attacked or defended. "I noticed that there was a lot of back-and-forth on the discussion pages," he says, "a lot of give-and-take very similar to what we were doing in the intelligence community debating various topics." And it was all happening out in the open, where everybody could benefit from the information.

Before the terrorist attacks on New York and Washington on September 11, 2001, the nation's intelligence agencies had been famous for keeping secrets to themselves. In the aftermath of that tragedy, as government leaders looked for ways to prevent another attack, there was widespread agreement that the agencies needed to share what they knew with one another. Intellipedia was provisionally accepted by top officials as one way to do that.

Set up on a trial basis in 2005 under the auspices of the Director of National Intelligence's office, the Intellipedia project has since expanded to include a whole suite of software tools that are now used by some hundred thousand individuals on three security levels: nonclassified, secret, and top secret. The busiest of these tools, so far, is the original wiki-based software, which attracts five thousand or so entries and edits a day. Like its enormously successful model, Wikipedia, which was launched in 2001 and now has more than 5 million articles, it functions through a process of *indirect collaboration* that resembles what we see in smart swarms in nature.

When someone comes up with an idea for a new entry in

Wikipedia, for example, he or she creates a short article called a stub, which is often no more than a placeholder for a full article. Consider the original stub for the term *stigmergy*, which was started on April 18, 2002, by a person identified only as 66.8.233.108 (an Internet Protocol, or IP, address). The stub defines stigmergy as "communicating indirectly by modifying the environment. Example: ants laying down pheromone trails." That was a good beginning, but not very informative. Since then at least sixty-five people have added bits and pieces to the article, pasting on sections about the history of the idea, its applications, references for further reading, links to related Web pages, or simply tweaking the text for clarity or punctuation. On July 15, 2004, someone identified as 63.146.169.129 noted that the Internet itself was a product of stigmergy, since users communicate with each other by modifying their shared virtual environment. "This wiki is a perfect example!" the person wrote. "The massive structure of information available here could be compared to a termite nest; one initial user leaves a seed of an idea (a mudball), which attracts other users who then build upon and modify this initial concept eventually constructing an elaborate structure of connected thoughts."

Like every termite who ever started a new mound, the individuals who launched the Wikipedia article on stigmergy couldn't have told you precisely how it would grow and develop over time. That job belonged to those who came along later, noticed the article-in-progress, and were inspired to improve it. That was one of the secrets of Wikipedia's success: *indirect collaboration* made it easy for almost anyone to get involved. "Many more people are willing to make a bad article better than are willing to start a good article from scratch," writes Clay Shirky, an expert on new media, in his book, *Here Comes Everybody: The Power of Organizing Without Organizations*. "A Wikipedia article is a process, not a product, and as a result, it is never finished."

As a model for intelligence analysis, this was the exact opposite of how agencies were accustomed to working, says Dennehy. "For a long time, the only way to get anything out of the government was through a structured vetting process that produced something with a seal on it that said this is our official position," he told the audience in Boston. "The process was edit, then publish. And the editing occurred in private." But Intellipedia and the other network tools flipped this model on its head, changing the order of the process to publish, then edit. "You get your initial thoughts out there," he says. "You never say it's done. What you're saying is, here's what I have now. Do you have anything to add?"

A good example of how this took place was described by Thomas Fingar, chairman of the National Intelligence Council, at a 2007 symposium in Chicago. The issue concerned the discovery by intelligence officers and others in Iraq that insurgents had started to use chlorine in improvised bombs. "Somebody said, we need a collection requirement for information on the use of chlorine in IEDs," Fingar says. A stub was created on Intellipedia, and it quickly attracted the attention of a number of individuals with useful information. "Over a period of about three days, if my memory is right, twenty-three people—some collectors, some analysts scattered around the world—put together, using Intellipedia, a perfectly respectable collection directive—what was needed," Fingar says. "And it just happened. Nobody said, use this tool."

A second important element in the Intellipedia suite of tools, in addition to the wiki software, were the blogs that analysts posted on the Intelink network. "A blog is a journal or diary that is kept in the public space of the Internet," Calvin Andrus had explained in his seminal paper. "Most blogs take the form of citing a current event and offering a point of view about it. Often one blog will cite a comment in another blog and comment on it. The 'blogosphere' is truly a marketplace of ideas." He suggested that those in the intelligence community who aren't used to seeing interpretation conducted quite so openly may be uncomfortable with this concept. For every brilliant idea expressed in a blog, he admitted, there were countless mediocre ones. But the trade-off could be worth it, because "as individual blogs comment on each other's ideas, the brilliant ideas will spread as feedback throughout the community," he argued. "Individuals, recognizing the brilliance, will respond. From this self-organized response will emerge the adaptive behavior required of the Intelligence Community."

The third tool in the Intellipedia suite was something called "Tag Connect," which was a customized bookmarking function like Delicious or Digg, two popular websites for the general public. If someone in the intelligence community found something on the Internet or Intranets that related to the plane crash in New York, for example, they could "tag" it with a bookmark, which was then organized with other bookmarks and made available to others in the community. "Social tagging gets bookmarks off personal platforms and enables other people to follow the tags," says Sean Dennehy.

Still other important tools were a photo management program called Gallery, which allowed intelligence personnel to upload and share pictures with one another, similar to what the public does through websites such as Flickr or Kodak Gallery; a video platform called iVideo that functioned like YouTube to let people share videos; an instant-messaging application called Jabber; a shared drive for documents called Inteldocs, which analysts could search using Google; and, finally, a subscribing function that notified users when key websites were updated.

Taken together, this complete set of tools—from the wiki platform to blogs, social tagging, picture and video management, instant messaging, document storage, and subscription—represented a full embrace by Intellipedia of the information revolution known in business circles as Web 2.0. This term, coined by new media guru Tim O'Reilly, refers to software and websites that run primarily on content generated by users, rather than on traditional materials published by a company or a news source. The photos on Flickr, for example, come from amateurs, not professionals. The videos on YouTube are often homemade. The basic idea, O'Reilly said, was that Web 2.0 organizations "embraced the power of the Web to harness collective intelligence," which was an attractive idea for many businesses. Of two thousand or so executives polled by McKinsey & Company in 2007, three-fourths said they planned to increase their use of Web 2.0 tools such as wikis, blogs, and social networking "to foster collaboration within the firm, to improve relations with customers, or tap the expertise of suppliers," among other goals. "The Web is no longer about idly surfing and passively reading, listening, or watching," write Don Tapscott and Anthony D. Williams in their book, *Wikinomics: How Mass Collaboration Changes Everything.* "It's about peering: sharing, socializing, collaborating, and, most of all, creating within loosely connected communities."

What Intellipedia had tapped into, in other words, was a broad enthusiasm for distributed collaboration, not just in the intelligence community, but in society at large. As Clay Shirky points out, "We now have communications tools that are flexible enough to match our social capabilities." These tools have given individuals the power not only to share things like photos, videos, and opinions, but also to form ad hoc groups to collaborate on common problems and to push for collective action.

Will Intellipedia revolutionize the way that intelligence is gathered and interpreted in this country? After its first four years in operation, the verdict is still out. Although it has proved its effectiveness at pulling together information from disparate sources, it has not yet gained wide use as a platform for collaborative decision-making, its supporters say. No one thought it would be easy, after all, to transform highly structured bureaucracies like the nation's intelligence agencies into the kind of flexible, adaptive, and self-organizing systems that Calvin Andrus first proposed. "The hardest thing in the world is to let go of control," says the CIA's Don Burke. "We have developed a society of control. Organizations are all about control. Society is all about control. These tools are about going organic."

As a shared platform for intelligence, however, Intellipedia has already demonstrated that the benefits of *indirect collaboration* are not reserved only for natural systems like ant colonies or termite mounds. By harnessing the power of stigmergy—in which a contribution by one individual changes a group project in a way that stimulates contributions by others—Intellipedia has given the intelligence community a better means to keep up with a rapidly changing world.

Which is what they did on the day of Cory Lidle's plane crash in 2006. As analysts using Intellipedia quickly determined by sharing bits of information, Lidle and Tyler Stanger had taken off from Teterboro Airport in northern New Jersey at 2:39 that afternoon, on the first leg of a trip to Southern California, where Lidle and his family lived. But before they left town, they'd apparently decided to take a brief sightseeing tour of Manhattan. After seven minutes in the air, they'd banked left around the Statue of Liberty and headed northeast past Governors Island up the East River, according to radar data from JFK Airport. They were flying at 700 feet one mile north of the Queensboro Bridge when they took a sharp left turn to avoid entering a restricted zone near La Guardia Airport. The river is 2,100 feet wide at that spot, but they'd been flying down the middle of the East Channel, which gave them only 1,400 feet for the turn. They were still making that tight turn when their plane struck the Belaire. To one witness, who reported seeing the wings "wobbling," it looked like the pilot was fighting for control.

No one knows exactly what happened in the cockpit during those last few moments, or why the pair had flown so far up the

East River. When asked by reporters, a fellow ball player speculated that Lidle had probably just wanted to see Yankee Stadium one more time before the season was over.

It's a Small World, Kevin Bacon

One January evening not long ago, thirty-eight undergraduates plopped down in front of computer workstations in a classroom in Levine Hall at the University of Pennsylvania to play a kind of game. Small dividers separated them so they couldn't peek at one another. They were told not to talk. On their computer screens, each student saw a small oval labeled YOU. Radiating out from the oval like the spokes of a wheel were lines connecting the oval to a number of small circles. These circles represented neighbors. They were red, green, or blue. The object of the game, the students were told, was to choose a color for their oval (from among red, green, or blue) that none of their neighbors had. The game would last five minutes, or until everyone had solved the problem. If, at the end of the game, every player was a different color from his neighbors, everyone would be paid five dollars. Otherwise, they'd all get nothing.

Put yourself in their place for a moment. Let's say that, at the start of the game, you have four neighbors. Three of them are green and one is red. What color do you pick for yourself? That's easy: if you don't want a conflict with anyone else, you pick the unused color, blue. Now let's imagine that, a few seconds after you choose blue, two of your neighbors also change to blue. They might have conflicts with people in other neighborhoods that you can't see on your computer screen. Now what do you do? If you change to red or green, you'll be in conflict with somebody else. But maybe that person has wiggle room in his or her neighborhood that you don't have in yours. So you change to green and hope that, as soon as that person sees the problem, he or she will fix it. And so it goes, move after move, until time runs out or the puzzle is solved.

If this sounds difficult, it is. In a formal sense, as mathematicians say, the Optimal Coloring Problem is "intractable" even for a computer. Yet the students were able to solve it (and collect their five bucks) in thirty-one of thirty-eight attempts, often in less than a minute. "I was pleasantly surprised by how well they did," says Michael Kearns, the computer scientist who organized and ran the experiments. "If I'd given you the same exercise this morning to solve by yourself, I'm quite certain you'd still be working on it this afternoon."

For a single individual, he explained, solving a coloring problem with thirty-eight circles is frustratingly difficult because every choice constrains what you do next. After you color in a certain number of circles, such that none conflicts with any other, you eventually just get stuck. "You're stuck but the mistake you made was way back at the beginning and it just took you this long to discover it," Kearns says. "The vague analogy I like to use is that it's like a maze. You could be just on the other side of the hedge from the exit, but you can't get out, because, back at the entrance, you should have gone left instead of right."

By contrast, each student in his classroom didn't need to solve the entire puzzle. In most cases, students weren't shown a bird's-eye view of the whole thing, only his or her own neighborhood. In that sense, the problem was distributed among the students in the same way that a colony of termites spreads the task of mound repair among workers. What made it all work was that each student's neighborhood overlapped with others, so that solutions made in one neighborhood rippled through the rest. Although it wasn't obvious to the students, as they interacted with others on their computer screens, they were all part of a web that connected them, in one way or another, to everyone else in the classroom. They didn't just belong to a neighborhood; they belonged to a *network,* not unlike, in principle, a group of friends on Facebook, intelligence analysts on Intellipedia, servers on the Internet, or power stations on the grid.

This was Michael Kearns's true interest. As someone fascinated by the role that networks play in collective intelligence—especially in situations such as games, where strategic decisions are being made—Kearns wanted to see how a group's behavior might be affected by the structure of the network that connects its members. Just as the structure of a termite nest affects the behavior of individual termites, shaping their experience of the world, so do the networks that we all use, from physical ones like the Internet to social ones like clubs or political parties. In a network, individuals interact with one another in different patterns and arrangements. During the past twenty years or so, scientists in a number of fields have studied the strengths and weaknesses of such patterns, which they depict in diagrams or maps, like the circles and lines on the students' computers. Some of these networks look orderly, like a family tree or corporate organizational chart, while others appear random, like bubbles in a boiling pot or a carton of buttons spilled on the floor. What effect do such patterns have on the functioning of these networks? Do some structures make tasks easier? Do others make them more difficult?

Consider the coloring game, in which individuals must distinguish their preferences from those of others. The problem was similar, Kearns suggested, to that of a teenager who wants a ringtone for her new mobile phone that none of her friends has. If the number of ringtones available is limited, how can she pick one not being used by anyone else in her network of friends? The solution to a problem like this depends on a give-and-take among several parties. What one person does affects the behavior of others. How are such interactions affected by the structure of the group's network?

To find out, Kearns linked students in his coloring experiments using two common types of networks. The first were so-called small-world networks, in which a surprisingly short connection can be found between any two "nodes," the term used for people or things in the network. One example would be the web of Hollywood actors that gave rise to the parlor game known as "Six Degrees of Kevin Bacon." The point of the game is to name an actor, and then see how many steps it takes to link that person to Kevin Bacon, who has a reputation for having been in a lot of movies. If you were to pick comedian Will Ferrell, for example, you might remember that Ferrell was a voice actor in the movie *Curious George* with actor Clint Howard, who was in *Frost/Nixon* with Kevin Bacon, giving Ferrell a "Bacon number" of two, since it takes only two steps to connect him to Bacon. The same task would be slightly more difficult with actress Amy Adams, who has a Bacon number of three. She was in the film *Dream Trap* with Charmin Talbert, who was in *Parenthood* with Rance Howard, who was in *Frost/Nixon* with Kevin Bacon. The fundamental insight conveyed by this exercise is that almost any actor can be linked to almost any other, not just Kevin Bacon, in relatively few steps, since the network of people who have appeared in movies is large enough (more than half a million) and interconnected enough to allow such short links.

Not surprisingly, the same insight may be applied to human society in general, as playwright John Guare suggested in his 1991 drama, *Six Degrees of Separation*. "I read somewhere that everybody on this planet is separated by only six other people," his character Ouisa Kittredge says in a well-known monologue. "Six degrees of separation. Between us and everybody else on the planet. The president of the United States. A gondolier in Venice." As unlikely as this claim sounds, a little reflection reveals how it could be true. If you can write down the names of fifty friends, relatives, coworkers, and neighbors, and each of them can do the same, then you're already connected to a network of 2,500 people in only one step. Expand the network by another step and you have 125,000 acquaintances, and so on, in a version of the same snowballing mathematics that we saw in the decision-making process of the honeybee scouts in Chapter 2. By the time you get to six degrees of separation, your network would include everyone on the planet. So if you live in Virginia, as I do, and you want to obtain some information about surfing in Coogee Bay, Australia, you might get in touch with your nephew Warren, who lives in San Francisco but used to live in Coogee Bay years ago and still has surfer friends there. In just two steps you have gained access to a dense social network on the other side of the planet.

The downside of this phenomenon, of course, is that it applies equally well to the sharing of swine flu as it does to e-mail addresses. In a network like the U.S. power grid, as we saw earlier, a cascading failure has no trouble finding a short path between a tree strike in northern Ohio and the billboards of Times Square, just as a financial crisis on Wall Street wastes no time in shuttering factories in China. The combination of dense local networks with a few long-distance links makes it a snap to spread either the good or the bad in a small world.

The second type of network used by Kearns in his coloring experiments was that described by experts as "scale-free," since it lacks a common scale by which its nodes can be accurately characterized. Take the global air-traffic network, for example, which includes an extraordinarily wide range of airports of different sizes, from a handful of giant hubs such as Heathrow, Hong Kong, and JFK, which have hundreds of connections to other destinations, to a great many smaller airports, such as Arctic Village, Alaska, or Taos, New Mexico, with links to only a few other places. If you were to ask how many cities you could fly to from an "average-size" airport in the world, you could not come up with a good answer, since there would be so few airports that actually matched the average.

The World Wide Web is another example of such a network, since it includes sites such as Google, Yahoo!, or Amazon, which link to countless other sites and serve as massive hubs for Web traffic, as well as many more low-traffic sites, which link to only a handful of other places. Because of this discrepancy in scale, the hubs are far more visible to users like you and me, which shapes our experience of the Web, and this only increases their popularity in a "rich get richer" way. In his book *Linked: The New Science of Networks: How Everything Is Connected to Everything Else and What It Means for Business, Science, and Everyday Life*, network scientist Albert-László Barabási describes this pattern of growth as "preferential attachment," saying that, when individuals need to choose between two websites, one with twice as many links as the other, "about twice as many people link to the more connected page." The more popular a Web page is, that is, the more popular it becomes.

Something similar happens in our relationships with friends and acquaintances. In his book *The Tipping Point*, Malcolm Gladwell describes a class of people he calls Connectors, who play a critical role in the spread of fads, fashions, and other trends through their participation in a wide variety of social circles. "All of us know someone like this," he writes. "They are the kinds of people who know everyone." We rely on such individuals to keep us in touch with what others are saying and doing, which makes them living hubs for gossip and word-of-mouth advice. "Perhaps one of the reasons why so many fashion trends don't make it into mainstream America is that simply, by sheerest bad fortune, they never happen to meet the approval of a Connector along the way," Gladwell writes.

One strength of such networks, the experts say, is that they are surprisingly resilient, with normal wear and tear. Random failures don't affect scale-free networks as much as some other types, since the vast majority of their nodes or components aren't linked to very many others. If an ordinary connection fails, it doesn't damage the network's overall performance. The failure of a hub, on the other hand, whether by accident or sabotage, can have serious consequences, since they handle so much traffic. You don't want Google to go down while you're researching a term paper, just as you pray that a snowstorm doesn't shut down JFK Airport while you're traveling. If either of those things happens, you may be in for a long day. At the same time, the vulnerability of network hubs to deliberate attack may also make it possible some day to prevent the spread of viruses like HIV (which causes AIDS) or bird flu by strategically targeting hubs in social networks. Vaccination of high-risk groups, such as health-care workers in the case of flu, may keep epidemics from spreading to the general population.

To compare the influences of these two types of network—small-world and scale-free—on the way students played his games, Kearns arranged for them occupy positions in each type, one after the other. The students didn't know which type they belonged to, since they were shown only their neighborhoods, not the whole network. Some were given three neighbors, while others (those playing positions as hubs) interacted with as many as twenty-two. No matter what position they occupied, as Kearns discovered, their behavior was strongly affected by the network's structure—but not in ways he necessarily would have predicted.

Scale-free networks, for example, made it much more difficult for students to win the coloring game than did small-world networks. Of seven experiments that ended without success, six were on scale-free networks. Even when students were able to solve the coloring problem on scale-free networks, they took longer to do so. Which makes sense, Kearns says, when you consider that the goal of the coloring game is to avoid conflicts with others, which is harder when you're highly connected. The most popular person in high school will always have more trouble choosing a unique cell-phone ring than someone who has only three friends.

"Think about what these hubs are good for," Kearns says, referring to the busiest positions in scale-free networks. "If the goal is to broadcast information quickly, they're very handy, because adding more links increases the speed of communication. If, on the other hand, you're trying to distinguish yourself from others, then you're going to have a really hard time, because you'll have a lot more conflicts. In that case, adding links just makes the problem harder."

This became clear after a second series of experiments in which the goal was the opposite of the coloring game. This time, instead of trying to be different from their neighbors, students were asked to be the same. Given a choice of red or blue, students were told they'd get paid only if *everybody* chose the same color by the end of the game. Their aim, in other words, was to seek consensus rather than uniqueness, which, if the pattern held true, should produce results that were the opposite of the coloring game. Being part of a scale-free network, in this case, should make it easier, not harder, for students to succeed.

To make the experiments more interesting, Kearns told a small number of students taking part that they'd be paid a bonus if the group as a whole selected red, while the majority of students were encouraged to choose blue. He didn't pick red players at random, though. He selected those with the most neighbors in the network, those occupying hub positions. Because of their role as connectors, he figured, they'd have a greater opportunity to be influential. Could such a well-connected minority impose its will on a less-motivated majority?

He didn't tell the students at the time, but this version of the experiment was directly inspired by the Democratic presidential primary season of 2008, which had dragged on for months without producing a consensus candidate. "It went on for so long that, at least for the first time I've ever seen, Democrats were actually saying, you know, this isn't good," Kearns says. "McCain is getting his campaign up and running and here we are, fighting with each other. We're looking bad." Even so, nobody wanted to tell the main contenders they shouldn't run. So the whole thing kept grinding along.

What would happen, Kearns wondered, if one or another of the candidates was able to gain the support of key connectors in the Party? If a small but influential group of individuals put their minds to it, could they win the nomination for their candidate? The answer, based on his experiments, was an unequivocal yes.

In 20 of 24 games, the minority was able to impose its preference on the majority—even when they were outnumbered six to one. Being highly connected in the consensus game turned out to be a powerful advantage.

The broader lesson, Kearns says, is that it doesn't make much sense to discuss the properties of network structures—such as small-world or scale-free types—without also discussing what it is you want to do on them. Different types of networks are better at different types of problems. Depending on the challenges an organization might be facing, for example, greater connectivity among workers or departments might be a good thing or a bad thing. Some organizations function best as a loose collection of tribes, each with its own specialists and experts, while others work better if they involve more collaboration. Tribal networks make it easier to differentiate one group from another, while greater connectivity makes it easier for an organization to achieve consensus. (At Boeing, as we saw in Chapter 2, Dennis O'Donoghue created a hybrid structure—in the form of the test operations center—to tap into the expertise of tribal departments such as mechanics and engineers, while also maintaining consensus about his division's common purpose.)

When we talk about networks, after all, what we're really talking about are relationships: how we interact with those around us (as well as those who interact with those around us, and so on), whether we happen to be supporters of a political candidate, marketers designing a TV ad, or power stations trying to keep up with demand on a hot summer day. When we take part in collective behavior, we're also taking part in networks, whose particular structures make it simpler or harder to achieve our goals, both as individuals and as groups.

It's easy to see how this works with termites, since their dense networks of passageways are physical expressions of interconnectedness. But other social insects such as ants and honeybees also rely on networks as they organize group efforts like foraging or nest building, with each task linked to others in dynamic ways. What one worker does affects what others do, just as each termite affects the behavior of other termites, and each student in Kearns's classroom affects the choices and actions of other students.

What we may not realize while we're taking part in networks is just how resilient they can be, even in the face of extreme challenges. That's what folks along the Gulf Coast discovered a few years ago when a storm named Katrina came calling.

Lost and Found

As it roared into southern Louisiana on August 29, 2005, Hurricane Katrina all but destroyed the region's communications systems—another of the critical networks at the heart of our modern society. Within hours, a thousand cell phone towers were knocked out, countless telephone poles were toppled, land lines were submerged, and switchboards were flooded, leaving more than 3 million people without phones for checking up on family members, reporting problems, or calling for help. It was a terrible time to be disconnected. All across New Orleans, once the levees gave way and most of the city was under water, tens of thousands of people were trapped on their roofs, in their attics, or in their hospital beds. At the Forest Towers East Apartments on Lake Forest Boulevard, a senior citizen residence in Orleans Parish, dozens of elderly men and women, three-fourths of them in wheelchairs, were stranded in the sweltering heat, with no way to escape or seek assistance. Like many others, they were on their own.

First responders, meanwhile, were struggling with their own network problems. Transmitters at police and fire departments had been destroyed, reducing radios to walkie-talkies with ranges of only a mile or two. With so many emergency personnel all sharing the same mutual-aid channel, officers on the street sometimes had to wait as long as twenty minutes to get a word in. Ham radio operators in other states were enlisted to relay messages to the handful of 911 centers that hadn't been flooded yet.

Government officials were cut off too. As Governor Haley Barbour later told Congress, the head of Mississippi's National Guard "might as well have been a Civil War general" for the first two or three days, because he could only find out what was going on by sending somebody out there. He did have helicopters instead of horses, so it was a little faster, he says. But not much.

The government's failure to anticipate such problems was only one part of a shocking collapse of central authority. Instead of providing a safety net for the region's citizens, those in charge demonstrated "confusion, delay, misdirection, inactivity, poor coordination, and lack of leadership at all levels of government," as a Senate committee later put it. While rescue helicopters remained grounded because of bureaucratic red tape, for example, some twenty thousand people languished in the stifling Superdome, where they'd been told to seek shelter. "First the air-conditioning failed. Then the lights. A generator kicked in, but with only enough power to keep the huge arena dimly lit," *Newsweek* reported. "The Salvation Army doled out thousands of ready-made meals, but bottled water was scarce, and in the steamy heat, the stench of unwashed bodies ripened. . . . On Wednesday, all running water shut off, and the reeking toilets overflowed." What happened to the government's promise to come to the rescue? critics demanded. Why did it take them so long to act?

In the midst of this colossal failure of centralized command and control, something unexpected happened. People began to improvise with forms of decentralized communication, using alternative networks to reach out to one another. Although voice lines failed, text messages were frequently getting through for those who knew how to use them. Messages were also popping up on Web pages, even those that weren't designed for such a function. "Please help," read one posted on the site of the *Times-Picayune* newspaper, which had temporarily ceased publication of its print edition. "About 400–500 people are gathered at the church at 5069 Willowbrook Drive, 70129 ZIP and are waiting for help. They have no food and water is rising. Please help." The woman who posted that note from a town in Texas was the daughter of a couple who knew people at the church.

More pleas for help appeared on other websites. "I really need to find my Dad," a young woman posted on the "Lost and Found" page of the New Orleans branch of Craigslist, a website normally dedicated to classified ads. She wasn't alone. Within days, thousands did the same thing. "Looking for Willie T. Samuels. His wife Sally is here in Little Rock, AK. Safe." Or "Bill Kovacs, I'm looking for you. Are you okay? Marci, in Miami." Or "Looking for my friend Lisa Porter. She is about 34/35 yrs old. Lives on Cypress Street. She's a smart girl, so I'm sure she got out, but can't find her and am very worried. I know of a few others looking for her too! Any information would be great!" As many as ten thousand posts were put up on Craigslist by people offering housing to Katrina victims. The Red Cross provided a "Safe and Well" list for displaced persons to register online so family members could catch up with one another.

As more and more people turned to the Web as their primary communications network, different kinds of sites sprang up. One was the Katrina Help Wiki, which collected information about shelters, relief aid, animal rescue, health and safety, and other issues. Created by volunteers who had worked on a similar site following the 2004 tsunami in Southeast Asia, the wiki soon attracted more than four thousand editors, who collaborated on data entry and other tasks."

No one expected the Web to be used this way in New Orleans. Its role as a lifeline emerged spontaneously from the wreckage of other communications networks, in the same way that natural swarms respond through decentralized behavior. When lives were put in jeopardy by the storm and the breakdown of government support, ordinary people pieced together new solutions, building upon each other's actions and reactions. They were just doing on the Web what they were doing in the rest of their lives: turning to the social networks they still had to salvage what they could from the storm's damage.

When experts describe such ad hoc responses to emergencies, they frequently use terms such as improvisation and adaptation. "Disasters may cripple a social system, but they do not completely destroy it," write disaster analysts Tricia Wachtendorf and James M. Kendra. "Indeed, the 'disorganization' that is sometimes seen after a disaster event is primarily the process of communities adapting to a newly emerging environment and new circumstances." When the existing social structure is damaged or destroyed, individuals or groups may collaborate in new ways that look messy but get the job done.

These same terms, on a smaller scale, could be applied to the distributed, adaptive responses of smart swarms. When Scott Turner and his team in Namibia knocked down that giant termite mound with a front-end loader, they set in motion a remarkable sequence of events, forcing hundreds of thousands of individuals, at a moment of emergency, to work together to recreate what had been lost. When they breached the wall of another mound, they stimulated a decentralized response from all levels of the termite colony. Something like that happened in New Orleans, says David Stephenson, an expert in emergency communication. "When you're faced with life and death, even the most traditional organizations are forced to abandon command and control tactics and explore innovative problem-solving."

The key lay in decentralized responses by private individuals. Those who survive a disaster must improvise solutions to keep up with changing circumstances, using whatever tools they have. "While the government failed miserably during Katrina, the people didn't," Stephenson says.

As an example of the spontaneous, decentralized actions that took place, consider the group that assembled in the Wal-Mart parking lot in Jennings, Louisiana, on Wednesday morning, August 31. With horror stories about the Superdome all over the TV news, three dozen citizens with seventeen boats responded to a call for help from Ronny Lovett, who owned a construction company near Lake Charles, about 175 miles west of New Orleans. Encouraged by his friends Sara Roberts and Andre Buisson, Lovett had asked employees with boats if any of them wanted to attempt what the government apparently couldn't get done—to rescue people trapped in flooded neighborhoods of New Orleans. "My guys don't back off of anything, and we all take care of each other," Lovett said to Douglas Brinkley, who told the group's story in his book *The Great Deluge*. "So, when Sara called, we all acted." Brinkley called this rough-and-ready flotilla the Cajun Navy.

Each boat was equipped with spotlights, chain saws, axes, life jackets, and whatever else they thought they might need, including a few firearms. "We had mechanics and electricians," says Roberts, a certified public accountant who helped organize the team. "Most of our crew members were skilled laborers, and we purposely assigned tasks according to their skills." Although these no-nonsense country boys were tough, many had never been to New Orleans in their whole lives. They weren't sure what to expect.

When their caravan of skiffs and hunting boats arrived at the staging area on the outskirts of the city, they discovered hundreds of other volunteers like them who'd also just appeared with their boats. An official with the Louisiana Department of Wildlife and Fisheries, which already had teams in the water looking for people, sent the Lake Charles group to another staging area near Harrah's Casino on Canal Street. There police officers from the New Orleans PD persuaded them that they should report to New Orleans East the next morning.

By seven a.m. on Thursday, Roberts and the rest of the Cajun Navy were ready to launch their skiffs down the flooded streets off Chef Menteur Highway in Orleans Parish. Before they set out, though, a NOPD officer explained the rules: "If you encounter a dead body, don't touch it. Leave it alone. That will be handled later. We are here to help people." There'd been reports of rescuers being shot at, he said. If that happened, they were to get out. "Bring those who will leave, but don't force them. No pets. Do not travel alone. Go out in pairs. As far as weapons, carry what you need for protection."

They rescued people all day, pulling them out of house after flooded house and ferrying them back to the Crystal Palace, a reception center normally used for weddings, conventions, and big parties. They were just doing what they could as part of an improvised, bottom-up solution to a massive problem. Then late that afternoon, coming down Read Boulevard, they spotted an elderly man waving an empty oxygen bottle out the window of a high-rise, desperately trying to get their attention. It was the Forest Towers East Apartments on Lake Forest Boulevard, on the next street over. The lost residents in wheelchairs had finally been found.

"We drove the boats into the dining room," Roberts told Brinkley, then carried scores of anxious residents down to their boats. "We had to get them to a stairwell where we had to get them through the dark water. In the stairway were mounds of dead rats, frogs, mice—it was disgusting."

A surprising number of residents were clutching pets. The NOPD officer had told them not to pick up animals, but Roberts decided then and there to ignore him. "I simply refused to tell these people who had been through so much that they couldn't take the only loved one they had," she says. "I lifted many a fat cat and put it into the boat."

Rules were important, Sara Roberts believed. But sometimes, in an emergency, you needed to make up your own rules.

4

Birds of a Feather

Secrets of Flocks, Schools, and Herds

A few days before Christmas in 1919, Edmund Selous was observing a flock of sparrows on a field in South Dorset overlooking the English Channel. The birds were doing something he'd marveled at many times before, something so puzzling it seemed to defy the laws of science. "Time after time," he wrote in his field diary, "a whole compact body darkening a considerable patch on the thin, close grassland, and numbering, I should say, at the very least some two or three hundred, but not improbably many more, whirled up instantaneously like some dark piece of stuff caught by the wind at all points." The sparrows, so far as he could tell, hadn't been startled into flight. He'd carefully kept his distance, and there was no one else around. He hadn't heard a gunshot or seen a dog running along the beach. Nor had the sparrows risen in a gradual wave, with one group following another. They'd left the ground all at once. One instant they were sitting or standing in the furrows, the next they were in the air. Almost as quickly, the birds wheeled in the sky and swept back to the ground, returning to practically the same area they'd just left. The whole thing seemed spontaneous and routine, Selous noted, as if it were an ordinary daily habit. "But an ordinary daily habit does not explain how hundreds can act together simultaneously as one," he wrote.

What puzzled Selous, who was a diligent observer and wrote a dozen books on natural history, was how so many individuals could coordinate their actions so precisely. Looking out on Weymouth Bay not far from his home, he'd often seen thousands of gulls rise up together as if "interconnected by some rigid arrangement of invisible wires." At other times he'd seen rooks, relatives of crows, do much the same thing, rising suddenly "with one animating impulse," then descending to another part of a meadow in a long, narrow black line. After watching a flock of starlings perform aerobatic maneuvers above the trees one evening, Selous described the way they changed direction as almost instantaneous. "Each mass of them turned, wheeled, reversed the order of their flight, changed in one shimmer from brown to gray, from dark to light, as though all the individuals composing them had been component parts of one individual organism." How was it possible, he wondered, for flocks to behave this way—not just to stick together but to flow as if they were a single being?

Selous rejected the theory that such synchronized behavior could be orchestrated by a leader, calling the idea "well-nigh unthinkable—it is *too* ridiculous." Nor could he accept the notion that the birds' spontaneous risings could be executed using the normal five senses. It all happened too fast. The only hypothesis he found plausible was that the birds were coordinating their actions through some form of mental telepathy, "some process of *thought-transference* so rapid as to amount practically to simultaneous collective thinking." If thousands of birds performed identical movements at the same instant of time, he figured, "must there not be in each one of them a precedent mental process—thought—also at the same instant of time?"

Studies of the paranormal were more common back then. In London, philosopher Henry Sidgwick of Cambridge had led research at the Society for Psychical Research to examine claims of human telepathy, clairvoyance, and other phenomena. In New York, psychologist William James, who cofounded the American branch of the same organization, was looking into hypnosis and psychic mediums for clues about the unconscious. "I cannot doubt that this class of functioning does exist," Selous wrote, referring to research published in the society's journal. "If it does, I for one should be inclined to look for its stronger and, as one may say, more normal exercise in nonhuman beings rather than in human ones."

His theory, to the degree that he spelled it out, was that birds shared a sort of pre-linguistic form of communication. "Their little minds must act together," he wrote. "Though I cannot understand it, yet it seems to me that they must think collectively, all at the same time, or at least in streaks or patches—a square yard or so of idea, a flash out of so many brains." Theirs was a different kind of cognition than ours, more primitive and diffuse.

Leaning on themes from Darwin, whose book *On the Origin of Species* had been published two years after he was born, Selous described telepathy in birds as "the imperfect calling back of something which we have lost." What, in humans, evolved into rational intercourse, he wrote, "may be with birds in numbers a general transfusion of thought in relation to one another."

Selous knew how unlikely all this sounded. But having spent decades in the field, he was confident in his observations, and telepathy struck him as the only theory that fit. Looking back on his writings now, it's easy to sympathize with his bewilderment. Clearly something was going on out there, with flocking birds, that resembled shared cognition. But what was it?

"Selous wasn't a crackpot," says Frank Heppner, a zoologist at the University of Rhode Island. "He was just a guy describing what he saw, very accurately. What he was saying, very simply, was there's something here that's unknown."

Today it seems clear that what Selous was struggling to understand was yet another form of decentralized, coordinated behavior. Just as a colony of ants can fine-tune its response to something new in the environment, so a flock of starlings can precisely adjust its shape and movement in the air. The key to such behavior is something I call *adaptive mimicking*, which, along with *self-organization*, *diversity of knowledge*, and *indirect collaboration*, is the fourth principle of a smart swarm.

By adaptive mimicking I mean the way that individuals in a group pay close attention to one another, picking up signals about where they're going and what they know. How they respond to such signals shapes the behavior of the group as a whole, which, in turn, informs the actions of individuals. As we've seen with ants, bees, and termites, the specific rules of thumb individuals follow in such situations still puzzle many scientists, not only those investigating social insects but also those studying flocks of birds, schools of fish, and herds of animals on land. What seems clear from recent research in biology, engineering, and the social sciences, however, is that the basic mechanisms of adaptive mimicking—*coordination*, *communication*, and *copying*—can unleash powerful waves of energy or awareness that race across a population, whether it's made up of starlings or caribou, stimulating the kind of coherent behavior we normally associate with centralized cognition. Their movements seem so organized that we can't believe they're not driven by a central intelligence.

What does this tell us about Selous's flocking birds? What's the secret behind their breathtaking displays in the sky? We know they can astound us with synchronized aerobatics.

But how do they do it?

Up on the Roof

The security guards were suspicious at first. They didn't understand why Andrea Cavagna needed to take so much equipment onto the roof of the museum. The Palazzo Massimo branch of the Museo Nazionale Romano in Rome holds priceless collections of ancient coins and jewelry as well as sculptures from the

Roman Empire. So the staff tends to be hard-nosed about special requests. But Cavagna was persuasive. The European Community had given him a grant for his research, he told them. They were counting on him to produce important scientific data from this project.

No one at the museum asked why Cavagna, who is a physicist, was studying birds in the first place. If they had thought about it for very long, they might have questioned his official government permit, which had been difficult enough to obtain. Nor did they wonder why all those bulky cases Cavagna and his team jammed into the museum's service elevator every afternoon contained mostly cameras and tripods, rather than scientific instruments. What were they doing up there, anyway?

"The truth is, we didn't know ourselves what we were getting into," Cavagna says of his part in the three-year project called Starlings in Flight (or STARFLAG), which ran from late 2004 to 2007. "If we'd known at the beginning how difficult it would be, we probably wouldn't have done it. We were kind of crazy, really."

Coordinated at the European level by Giorgio Parisi, a fellow physicist at Italy's National Institute for Condensed Matter Physics, the STARFLAG project involved teams of biologists, physicists, and computer scientists at seven research institutes in Italy, France, Germany, Hungary, and the Netherlands. Their goal was not only to gather empirical data about starlings in flight, but also to improve 3-D simulations of the birds and develop insights, if possible, into intriguing similarities between flocking and group behavior by humans, such as fads, fashions, and herding in financial markets. Cavagna's team was given the critical task of producing 3-D empirical data by photographing the flocks over Rome.

As a statistical physicist, Cavagna normally spends his time scribbling equations on a blackboard, not handling equipment in a lab. His area of expertise is disordered systems theory, a branch of physics that deals with complex systems in which many elements interact in unpredictable ways. Before the STARFLAG project began, Cavagna had practically no experience in organizing an experiment in the field. Yet in late November 2004, when funding from the EC came through, he purchased cameras, tripods, computers, and other gear and led his small team of graduate students to the roof of the museum.

Cavagna had picked the Palazzo Massimo building because of its location on a broad tree-lined plaza near the Termini railway station, the city's main terminal. Every afternoon during the winter, flocks of European starlings (*Sturnus vulgaris*) that have spent the day feeding in fields and groves outside Rome return to roost in the trees of city parks or plazas like this one, presumably because such spots are warmer than the surrounding countryside or offer a safe haven from predators. In one of the most beautiful and mysterious displays of coordinated behavior in nature, these flocks typically perform aerobatic maneuvers in the sky for a half-hour or more at dusk before settling down in the plaza's trees. Depending on the density of the birds in a flock, the group may look like a wispy column of smoke or a playful beach ball rising and falling in the air. Sometimes a flock will twist back on itself like a ribbon, showing a mesmerizing fluidity.

"I've seen the birds do this hundreds of times," Cavagna says, "but I still find it beautiful."

No one knows for sure why the starlings put on these amazing displays. Other birds exhibit impressive flocking behavior, from Canada geese with their orderly V-formations to pigeons that swirl over city streets, but none shares the exuberance of starlings. Some biologists have speculated that these performances function as flashing billboards to guide flocks to roosting spots. Others have proposed that they're a way for flocks to assess the strength of their own numbers, which might play a role in regulating breeding. Still others have suggested that they're training exercises to prepare flocks to evade predators. The more tightly packed a flock becomes, the thinking goes, the more difficult it should be for a hungry falcon to zero in on a single bird. To Cavagna and his team, who frequently observed flocks twisting and turning to escape from a pair of falcons that lived near the train terminal, the predator theory sounded logical as a defensive measure. But they agree that it doesn't really explain why the birds perform such displays every night.

"I mean, think about it," says biologist Frank Heppner. "The starlings tend to roost in the same place every night. The peregrine falcon isn't stupid. He sees these guys coming in every day. If you were a starling, wouldn't the prudent thing to do be to dive right into your roost, where you have some protection? Why would you spend forty-five minutes turning and wheeling over the roost, making yourself an available target for the peregrine? That doesn't add up to me."

Whatever the reason for the displays, Cavagna and his team

were trying to figure out *how* the starlings did them rather than *why*. From their vantage point on the roof, they were ideally positioned to document the birds' movements in three dimensions. The only disadvantage of the location for Cavagna was that the museum staff stubbornly insisted that they remove all their equipment every day, which meant they had to spend an hour and a half each afternoon remounting cameras on tripods, realigning the cameras, refocusing them, and resynchronizing their electronics. All of which introduced opportunities for errors.

"Our first season was a disaster, a real disaster," Cavagna says. "We made an amazing number of mistakes." After two months of trying, he and his team obtained usable images on only three days of observations. But they learned enough from these early efforts to improve their performance the following year. Because of technical issues, Cavagna had ended up purchasing six high-end digital cameras, Canon EOS Mark IIs, costing about $4,000 each. Although these still cameras weren't as fast as high-definition video cameras, they were capable of taking pictures with up to eight times more detail than standard video cameras. This was critical for gathering data, since images of birds that weren't sharp would be useless. To compensate for the cameras' slower speeds, Cavagna mounted them in pairs on rigid aluminum bars, then wired each pair to stagger the phase of their shutters. This effectively doubled their speeds from five to ten frames per second.

Since his goal was to create stereoscopic images, Cavagna positioned two of the tripods, each holding a pair of cameras, 80 feet apart on the museum's rooftop. Then he stretched forty-pound nylon fishing line between the tripods to precisely align the cameras' fields of view, focusing on a spot 330 feet away, where he estimated the flocks would fly. Sometimes he and his crew were lucky in this regard, capturing up to forty stereoscopic images in a three-second sequence, and sometimes they weren't, such as when a flock drifted out of the frame. Images from the third pair of cameras were used during the analysis phase to triangulate the positions of individual birds with each flock.

"We got better the second winter," Cavagna says. From November 2005 to February 2006, they recorded ten flocking events. Shooting almost every day for four months, they captured another twelve flocking sequences that were of high enough quality to be digitally reconstructed on their computers.

The primary question for Cavagna was how a flock maintained its cohesion during flight. How did several thousand birds stick together while executing the most extreme kinds of twists and turns? Why didn't they crash into one another or fly off in different directions? If this extreme *coordination* was the result of self-organized behavior—if individual birds were following simple rules to interact with one another as they flew—what could those rules possibly be? Previous efforts to document flocking in three dimensions had yielded modest results. The first stereoscopic images of starlings and dunlins, captured thirty years before by Canadian biologists near Vancouver International Airport, showed loosely organized flocks of no more than seventy birds. A similar study in the late 1990s in England using low-resolution video cameras documented even smaller flocks. When Frank Heppner and Harold Pomeroy set up an experiment with trained homing pigeons near Providence, Rhode Island, in 1980, their flocks averaged only a dozen or so birds—and proved frustratingly difficult to capture on film.

"It took us a long time to train them to do what we wanted them to do, which was to fly in a constricted area near the roost," Heppner says. "When the day finally came to load film and take our first actual images, we were very excited, because these would be the first real sequence pictures of birds in flight, meaning we could get flight paths from them. We knew the mathematics would work. So we let the pigeons out of the roost, and they headed over to our training area. I guess we had about nine birds, and they were turning and wheeling. And out of the woods, like a bat out of hell, comes a Cooper's hawk. It heads smack into the middle of the flock, which explodes, and all the birds just disappear. We never saw them again."

They had better luck with their second flock, which revealed some interesting new information about the way pigeons change positions during turns, but Heppner and Pomeroy were still taking baby steps in documenting the mysteries of flocking. Cavagna and his team were hoping for more. Unlike these earlier investigators, they wanted to document flocks as large as those that typically occur in the wild—from a few hundred to tens of thousands. Only by photographing big flocks, they believed, could they gain a true understanding of how individuals interacted. And besides, as physicists, they were more comfortable with data involving large numbers.

"To a biologist, a flock with relatively few birds was not a problem," Cavagna says. "But for physicists like us, it was just a tragedy. Because we believe that size matters, that big numbers do things different from small numbers. If we wanted to study emergent collective behavior, we knew we needed to have big numbers."

Big numbers meant big headaches, however, as soon as they started downloading images into their computers. From fifty acceptable flocking events, they selected ten sets of images for further analysis. Their plan was to convert pairs of images into 3-D reconstructions of the flock, assigning coordinates in three dimensions to every bird. But before they could do that, they needed to solve two thorny problems related to image analysis.

The first was fairly straightforward: How to tell one bird from another when several of them overlapped in a photograph? Heppner and Pomeroy had known their trained pigeons well enough to identify each one by its unique markings. But with thousands of identical-looking starlings in as many as eighty photographs per flocking event, separating them by hand quickly became impractical for Cavagna. So his group invented an algorithm called the "blobsplitter" to tease apart clumps of birds into as many as ten individuals by tweaking the intensity level of the pixels in each clump. Each bird was then assigned its own coordinates.

That led to the second problem: How to determine if a bird identified in one picture from a stereoscopic pair was the same as that seen in the other? Since photographs in a pair were taken from different positions in space—one from the right side and the other from the left—they depicted the same birds in very different relative configurations. From a practical point of view, matching a bird on the edge of a flock with its corresponding image was relatively easy. But doing so for the many birds crowding the middle turned out to be a nightmare.

"This was the problem that had blocked people in this field for thirty years," Cavagna says. His team solved it by adding the third pair of cameras. Their procedure, to oversimplify, went something like this: First, they applied a pattern recognition algorithm to fifty or so of the easiest matches, using a statistical technique that flagged those with the highest probability of being accurate. Next, they used the coordinates of those birds to precisely fix the geometrical relationship among the three cameras. Then they applied a relatively simple form of triangulation to solve the rest of the matches. "To develop this procedure took more than two years," Cavagna says. "But now we can take the original JPEG, the digital photograph, feed it into the program, and after some time, which can be ten minutes to two hours, depending on the size of the flock, we get the three-dimensional coordinates of all the birds."

After sorting out the locations of all the birds, Cavagna's team reassembled each flock into a 3-D image that they could rotate and study. What they discovered from these images surprised them: many of the flocks that had looked roughly spherical from the museum rooftop turned out to be remarkably flat. "We thought these flocks were elliptical in shape like a potato," Cavagna says, "but in fact they were more like potato chips." What's more, birds in many flocks were distributed in an unexpected way. Instead of being evenly spaced out like molecules of gas in a balloon, birds near the edges of the flock were more densely packed than those in the middle. "It was more like people on a bus," Cavagna says. "When I get on a really crowded bus, I always see people in the middle chatting leisurely while those of us near the door are completely squeezed. It could be the same with the starlings. Those on the outside of a flock push a little harder to get inside, where they feel safer from attack by the falcon."

But the most intriguing finding concerned the rule of thumb that birds seem to use to stick together during flight. Each bird, Cavagna's team discovered, interacts with relatively few neighbors, and those neighbors tend to be located on either side of the bird, rather than in front of or behind it. It doesn't matter if these neighbors are close or relatively far away. It's the *number* of neighbors that counts, not the distance between them, and that number, on average, falls between six and seven.

It's not that each starling can't see other birds. On average, a starling might find itself with fifteen or sixteen birds in its field of view. But each starling pays attention to only the nearest six or seven. These birds tend to be located on either side, which makes sense, Cavagna says, when you consider that a starling's eyes are located on the sides of its head, making it easier to look sideways than forward or backward. But how did Cavagna's team come up with the figure of six to seven?

This was where their training in statistical physics paid off, Cavagna says, because measuring the relative positions of particles in space was something they knew how to do with precision. "Imagine that you have a sphere with a seat at the center," he explains. "You take Bird One, and you put him on the seat, and you ask him, Where is your nearest neighbor? And he says, There. And you put a dot on the sphere. Okay, next: Bird Two. You put him there. Where is your nearest neighbor? There. And you put a dot. Then you have all these dots on the sphere. And then you say, What is the distribution of the dots? With the starlings you will not see an even distribution of the dots. They will be grouped on both sides of the birds, but very few will be in front of or behind them."

Physicists call this kind of distribution "anisotropic," which means that things are not evenly spaced out. The opposite would be "isotropic." Consider raindrops on a sidewalk, Cavagna says. "If you take one drop and you look around the drop, you might find the same distribution of drops in all directions. That would also be isotropic," he explained. "Or consider the sky at night, which is equally dark in all directions. Darkness is isotropic, too. But during the day the sky is not equally luminous in all directions, because the sun is very bright. In other parts of the sky, the light is weaker. So in that case the sky is anisotropic."

The fact that starlings tried to keep their nearest neighbors on either side of them meant that their relative distributions were anisotropic—unevenly spread out. But the researchers discovered that as they moved from the nearest neighbor to look at the second nearest, or third nearest, the anisotropy tended to weaken. By the time they got to the tenth or twentieth nearest neighbor, the strength of the interaction was completely gone. At that distance, birds stood an equal chance of being in any direction from the starling. The distribution was isotropic. The exact point where the interaction strength faded away—and the distribution shifted from anisotropic to isotropic—fell between six and seven.

The key point was that the strength of this interaction decayed as the number of neighbors increased, not as the distance to the neighbors increased. Previously, most computer simulations of flocking had assumed that, to maintain cohesion, individual birds coordinated their movements using a kind of zone system measured in physical distances.

The first simulation of this kind was a deceptively simple program called Boids, created in 1986 by Craig Reynolds, a computer graphics expert. In his program, Reynolds gave birdlike objects, or *boids*, three rules to follow: Avoid colliding with other boids, stay close to nearby boids, and fly in the general direction of these other boids. By adhering to these rules, the boids did a convincing imitation of flocking as they moved across a computer screen. "Each boid was trying to decide how to steer to accomplish those three things," Reynolds says. "If you were a boid, you'd steer away from a neighbor who was too close. You'd avoid local congestion, try to turn away from collisions with your neighbors, and align with them if possible. Then when you were farther away, you'd try to turn toward them. So there was a repulsive force and an attractive force in different range scales."

Reynolds set up a kind of bubble around each boid, which he described as its "neighborhood." If other individuals entered this neighborhood, then the boid responded to them. If they didn't, it ignored them. The size of this neighborhood was defined by a distance measured from the center of the boid and an angle measured from the boid's direction of flight. By following this relatively simple system, boids were able not only to stick together with their flock but also to dodge stationary obstacles, like pillars or buildings, in a lifelike manner.

Tamás Vicsek, a Hungarian physicist, created a similar model in the 1990s, inspired by the behavior of magnetic particles, which are known to align with one another under the right circumstances. Was it possible, he wondered, that flocking, schooling, and other biological phenomena were driven by similar principles of alignment? To find out, he set up a series of simulations in which a group of generic particles, all moving at the same velocity, were programmed to follow a single rule: Head in the same direction as the average direction of all the other particles within a certain radius—a zone similar to a boid's neighborhood. The results were fascinating to watch. As Vicsek increased the density of particles in each simulation—from a minimum of forty to a maximum of ten thousand—the particles spontaneously started clustering, at first in small groups that wandered in random directions, then later in a single group moving in the same direction. The mystery of flocking, his experiment seemed to say, was no more complicated than the science behind a child's toy magnet.

The big difference between such models and Cavagna's starlings, of course, was that the rules of interaction followed by the starlings weren't based on a zone system measured in physical distances. Instead, the starlings tracked their nearest six or seven neighbors, no matter how far apart they were. Such a system, Cavagna observes, was based on "topological" rather than metric distance, which gave a starling flock its remarkable elasticity.

"We found that interactions between a given bird and its nearest neighbors were just as strong when the birds were five meters apart as they were when they were one meter apart," says Cavagna. By tracking the same number of birds, regardless of whether a flock expanded or contracted, each starling was able to adjust to sudden changes in density—the kind a flock experiences when attacked by a falcon. Not surprisingly, according to Cavagna, evolution "has selected the interaction rule that is most resilient to external attacks."

But why did evolution pick six or seven as the magic numbers? Why not four or five? Or nine or ten? The answer wasn't clear to Cavagna based on his team's analysis of flock structure. But on a recent winter afternoon in Rome, he speculated about two possible explanations.

"One theory is that the number seven represents the limit of a bird's cognitive ability," Cavagna says as he stood on a bridge over the River Tiber near St. Peter's Basilica. Several groups of starlings had gathered in the sky, attracted by a line of horse chestnut trees along the river, where they intended to roost for the night. "I'm not talking about the ability to count, which is something that very few animals have, but something simpler—an ability to distinguish between different numbers."

During the early 1990s, Jacky Emmerton and Juan D. Delius, psychologists at Purdue University, had shown that pigeons could tell the difference between a circle containing two dots and one containing three and could give the right answer more than 70 percent of the time. The pigeons could even distinguish between six dots and seven. But when asked to choose between seven dots and eight, they couldn't do it, and they gave the wrong answer more than half of the time. "They just weren't able to keep track of more than seven objects," Cavagna says. "They got confused. Which could be a reason why starlings don't track more than seven of their neighbors."

Tens of thousands of starlings were swirling overhead now in diffuse clouds. "I call this the soup phase," Cavagna says. "They're all over the place. As it gets darker, they will start flocking in a more serious way. It's like they have an easygoing way of staying together and a no-jokes way." A steady rain of bird poop was falling on the bridge, prompting him to pull up the hood of his windbreaker.

"Okay, the second theory about the number seven is that it represents the optimal number to spread information across a flock," he says. "You might think that the more individuals you interact with, the better. But it's not necessarily so." If you interact with too many individuals, he explained, you get too much information. And a lot of that information will be "noisy," or inaccurate, since it comes from individuals that don't have a clue about what's going on. If you interact with a relatively small number of individuals, on the other hand, the information will be more reliable. But if something important happens, you might find out about it too late. If a falcon shows up, for example, and starts attacking your neighbors, you might not get the news until the falcon is very close to you, which would not be good. So there must be a number that optimizes these factors, and natural selection may have hardwired that number into the starlings' behavior. Simulations by other researchers have suggested that the ideal number might be three and four neighbors, Cavagna says. The question remains open.

By now, the chattering of the birds overhead had grown as loud as the city traffic. A street vendor at the end of the bridge tried in vain to scare the birds away from the trees above his food stand by slapping pieces of wood together. Tourists holding maps lifted up their faces to watch the show. Was it possible, in the end, that starlings performed these amazing displays because they were simply good at it, Cavagna wondered. What if they flew with such abandon because that's what they were built for? "You know, when we were children, in a way that's what playing was for, if you want to use that as a conceptual shortcut," he says. "The birds know how to do this amazing flocking, and they're uncomplicated enough to always do the same thing, even when they don't need to, because this is their nature. I'd say that makes perfect sense."

Whether their nightly displays are expressions of their joy at being birds, or something else, the starlings had revealed a big secret to the Italian team. The remarkable elasticity of their flocks—their extreme *coordination* during aerial maneuvers—was the result of simple interaction: by tracking their closest six or seven nearest neighbors, the birds had learned how to swerve to survive.

Something similar seems to be at work in many other groups, where imitation is used as a ready vehicle to disseminate influence or information. As we'll see with schools of fish, herds of caribou, and even groups of people, paying close attention to others can become a powerful source of collective intelligence when it provides members of a group with information they couldn't acquire by themselves. By shaping interactions among individuals, meanwhile, copying also lends structure to groups from the bottom up. Such groups, in a sense, have solved the problem of coordinated behavior as skillfully as ants have solved the problem of task allocation, or honeybees have solved group decision-making, or termites have solved problems of nest building.

This was a valuable insight for a special-effects wizard in New Zealand who set out to depict collective action in movies. If he wanted to create scenes as spectacular as those in nature, he realized, there was only one way to do it: by copying the strategy invented by birds.

Flocks of Movie Monsters

On Valentine's Day, 2004, in the ballroom of the Ritz-Carlton hotel in Pasadena, the actress Jennifer Garner presented an Academy Award to New Zealander Stephen Regelous for his work on the *Lord of the Rings* trilogy. Regelous wasn't being recognized for acting, directing, or screenwriting. He was being honored for populating the movies with swarms of lifelike monsters.

A former graphic designer who'd taught himself computer programming on the side, Regelous was thrilled. "I haven't won anything since a coloring competition when I was eleven years old," he says. He'd been hired by fellow Kiwi Peter Jackson, director of the *Lord of the Rings* films, to help create epic battle scenes with hundreds of thousands of computer-generated warriors. The digital extras he produced, including graceful Elves, hideous Orcs, and brutish Uruk-hai, had to be as convincing as possible, Jackson believed, to do justice to the world of Middle Earth as it had been imagined by J. R. R. Tolkien in his novels. So he'd encouraged Regelous to pursue a longtime dream: to create flocks of animated characters that could think and move on their own. His effort was so successful, movies haven't been quite the same since.

Consider the final battle scene in the second of the three films, *The Two Towers*. In it, the heroic Rohirrim, a race of humans, make a desperate last stand against the hideous forces of Saruman, a dark wizard, inside a medieval-style fortress wedged into a narrow canyon. The scene calls for an old-fashioned spectacle: several hundred knights in armor looking down from a stone parapet onto a sea of ten thousand subhuman warriors bearing torches and lances. As the forces of darkness march relentlessly toward the castle, they flow around a giant rock on which their monster commander defiantly stands, each warrior jostling and pushing forward. When they reach the foot of the fortress, the army raises ladders to scale the wall, with a sword-wielding warrior on the end of every ladder. The defenders shoot volleys of arrows down on their slobbering enemy, but there are too many of them. The monsters flow up the ladders like ants, and the sword-fighting begins in earnest inside the castle.

There's no way to tell from the final sequence that Jackson

hadn't hired ten thousand live extras for the battle. When the camera flies over the parapet and scans the approaching horde, a flash of lightning reveals an army stretching to the horizon. In 2002, when *The Two Towers* was released, of course, audiences were accustomed to seeing vast crowds generated by special effects. But this was no ordinary movie trick, like those where groups of two hundred actors were digitally cut-and-pasted into a landscape over and over again. This was something new. In a way, Jackson *had* hired extras, only they were virtual ones rather than flesh-and-blood.

Each warrior in Saruman's army—despite not being real—was moving on his own, guided by what he saw, heard, and felt during the battle. Each moved at his own pace, carried his lance in his own style, and was careful to keep from bumping into other warriors, even as they flowed around the giant rock. When the ladders went up, the warriors on the end of each one were virtual actors that Regelous calls "agents." When others followed up the ladders, nobody, not even the movie's director, was telling them what to do. They attacked knights in the castle in their own way.

"When I say they have vision and hearing and a sense of touch, I'm not talking metaphorically," Regelous explained. "They really do have vision, in that, on every frame of the movie, for every agent, an image was created by rendering the scene from their point of view. And those pixels actually got fed into the agent's brain, where rules were applied to the pixels in that image."

If an agent were programmed to fight, for example, the character might respond to an enemy nearby by turning toward him and swinging his sword. If he heard a loud noise, like an explosion, he'd probably look in that direction, like all the other agents around him. There was no telling precisely how he might behave, since his fuzzy-logic brain could respond in an almost infinite number of ways to the situation he was in, which made his actions seem only more lifelike.

"When we did one of our tests for *Lord of the Rings*, we had a group of Elves encountering a horde of Orcs, and the Orcs, not being as bright as the Elves, just rushed toward them," Regelous says. "So the Elves formed a horseshoe around the Orcs to surround them. Now, we kind of forced the Elves to do that by having them follow a particular color on the terrain. But one thing we didn't expect was that, when they pushed forward to form this horseshoe, some of them couldn't see properly as they moved toward the inner edge of the horseshoe, where they would fight with the Orcs. So they actually stepped around a bit and reshuffled themselves until they got a better vantage point. And we didn't put in any rules for them to do that. It was just an emergent property of the very simple rules that we'd given them to approach the Orcs in this horseshoe shape. But it looked really cool. It was like, wow, look what that guy's doing!"

In another early test, Regelous pitted a group of a thousand generic sword-fighting agents against a second group of the same size. "Most of them engaged in battle. But in the back you could see a couple dozen guys running for the hills. It was like, yup, they're running away. They're not stupid. They don't want to fight." As it turned out, it wasn't quite that simple. What the fighters lacked wasn't courage but a sense of where the enemy could be found. As soon as Regelous added some rules to make them turn around if they couldn't see any enemy, they charged back into the fray.

To learn how such animated groups ought to behave, Regelous had studied similar groups in nature, including flocks of birds and crowds of people. From this research sprang his "bottom-up" approach to simulation, in which characters interacted with one another like starlings in the sky or pedestrians in a train terminal. "The whole point is, to model a crowd, you model lots of individuals, and the crowd is an emergent property of them reacting together," he explains.

Just like a flock of starlings, in other words, the monstrous Orcs in *The Lord of the Rings* coordinated their movements as a marauding army by following simple rules of interaction: stay close to other Orcs, don't bump into other Orcs, head in the same direction as the Orc throng, and, if you happen to run into any humans, cut them in half with your sword.

In the years since the *Lord of the Rings* movies, the company Regelous founded, Massive Software, has sold animation systems for use in dozens of films, commercials, and computer games to simulate not only crowds of people but also groups of animals. In the 2006 comedy *Night at the Museum*, for example, it was used by filmmakers to stage a chaotic battle between two-inch-tall occupants of diorama exhibits, including cowboys and Indians, Roman centurions, Mayan warriors, and Chinese railway workers. In *The Ant Bully*, an animated film in which a boy is reduced to the size of an ant, the software helped depict forager ants as they followed a pheromone trail. And in the movie *Evan Almighty*, the software populated the ark built by Steve Carell's Noah-like character with thousands of animals, including flocks of birds.

"It's very easy to do birds in Massive," Regelous says. "It's not really built into the software, but it's easy to make a brain with the basic rules of flocking. That's very much along the lines of the emergent properties we see in nature that can be modeled with simple rules in individuals." In fact, a Spanish ad agency and a special effects studio not long ago teamed up to create a TV commercial for Honda that combined images of real starlings with a computer-generated flock, using software from Massive.

Many of the television commercials made with Massive software feature large groups of people in the background, such as the Verizon Network team that follows cell-phone subscribers around, or the stadium crowd in the Budweiser ad that does the wave to pour beer from a giant picture of a bottle into a giant picture of an empty glass. To create that ad, an agency called Method Studios populated a stadium with some 97,000 animated characters, using a ready-to-run agent capable of simulating fans doing a wide range of things, from quietly watching the game to standing and cheering, clapping, talking on cell phones, having conversations with one another, or even doing the wave.

More recently, Regelous and his company have ventured into real-world applications of his virtual agents. An engineering firm in Los Angeles, for example, has experimented with a new Massive program designed to simulate how occupants of real buildings evacuate during a fire, hoping it could show architects how to design safer structures. Because of the software's unique capabilities, including an agent's vision, hearing, and touch, as well as memory of where it's been, the engineers think Massive can provide a richer portrayal of human behavior than the kinds of models we saw earlier in this chapter, in which birds were depicted as if they had zones of awareness to keep them from colliding. What have the engineers learned so far from the Massive simulations? That people who ignore fire alarms, for one thing, are more likely to get up and go if they see other people heading for an exit.

The ultimate application of Massive software, though, might be to simulate human behavior through robotics. At the Wired Nextfest technology trade show in Los Angeles in 2007, a small Dallas-based company, Hanson Robotics, introduced a seventeen-inch-tall robotic boy named Zeno, whose brain was essentially run by Massive software. With spiky black hair and a flexible, expressive face modeled after the son of inventor David Hanson, Zeno smiled, blinked, shrugged his shoulders, looked determined or surprised, all in response to things he saw or heard. Because of the "character engine" that Hanson had built into Zeno, the boy robot was able to hold amusing conversations with people at the show, make eye contact, and even remember people's faces. In fact, according to the latest facial recognition tests, Hanson told one reporter at the show, "Zeno recognizes faces better than people do."

Hanson's company says it hopes to sell the first Zenos in a few years for about $1,500 each as educational toys for children. The goal, according to the inventor, is to help Zeno continue to learn and grow more aware over time. For Massive Software, meanwhile, Zeno's debut seems like the completion of a circle. What began as an ambitious project by Regelous to simulate the unpredictable behavior of individuals in a virtual world has taken physical form as a boy robot.

Training a Team of Robots

In the 2002 science fiction film *Minority Report*, a discredited chief detective named John Anderton, played by Tom Cruise, is hiding from authorities in a tenement in Washington, D.C. As members of a SWAT-like squad prepare to search the building, two officers unclip eight small devices from their belts and toss them on the sidewalk, where they spring to life as robots called "spyders." Swift and agile, these tiny reconnaissance units race up the front steps on legs as long and elegant as those on a daddy longlegs.

Working as a team, the spyders scurry from floor to floor, entering apartments by squeezing under their front doors. Once inside, they corner residents, climb up onto their faces, and scan the irises of their eyes to determine their identities. They're systematically checking out all twenty-seven of the warm bodies who appear on a real-time thermal image of the building taken by the police. But when the spyders get to John Anderton's apartment, they find the door blocked by towels stuffed underneath. As several scratch at the door, one turns around and explores the floor of the hallway, discovering a panel that leads to a duct. It opens the panel, and the rest of the robots scramble into the duct, following it into Anderton's apartment.

Anderton, meanwhile, has filled a bathtub with cold water, tossing a cooler's worth of ice cubes into the water for good measure. He climbs into the tub, takes a deep breath, and sinks to the bottom, hoping to evade detection. He's still holding his breath when one of the little three-legged robots enters the bathroom, sweeping its searchlight from side to side the same way an ant toggles its head while foraging. Finding nothing suspicious, the spyder is about to leave when a tiny bubble escapes from Anderton's nose, rises to the surface, and pops. In a flash, the spyder races back into the room, accompanied by the others, with whom it shares instantaneous communication. They surround the tub, climb up onto its rim, and peer down into the water. One of them reaches a leg into the water and gives it a jolt of electricity, which pops Anderton to the surface. The rest join in on the zapping until he submits to an eye scan—like everyone else.

It's a dark vision of the future and the role that technology could someday play. Despite enjoying the benefits of virtual touch-screen computer interfaces, shopping mall billboards that address them by name, and cars that drive themselves, people in the year 2054 don't look much happier or more prosperous than they do today. They certainly haven't had much success improving life for the poor. As the spyders race down the tenement's hallways, they pass chunky rats scurrying the other way.

There's something both familiar and disturbing about these robots. In appearance, they resemble insects, but their cunning reminds us more of a pack of lions or wolves. Such behavior represents one end of a spectrum of capabilities—from carrying out search-and-rescue missions after a natural disaster to conducting search-and-destroy missions during battle—that scientists have imagined for robot teams in the not-too-distant future. More often than not, the inspiration for such robots has come from swarms of ants or flocks of birds.

Consider the scene at Fort A. P. Hill in Virginia on January 20, 2004. A group of scientists from SRI International of Menlo Park, California, under contract to the Defense Advanced Research Projects Agency (DARPA), released a platoon of sixty-six foot-long robots into an empty office building. Their mission: search for something hidden inside. As they wobbled down the hallway on three wheels, the little red robots looked both cute and unnerving. About a foot tall with round bases, these units were equipped with Web cameras for surveillance, eight sonars for navigation, and wireless communications gear to share information with one another. Sometimes, one of them would stop in its tracks and slowly rotate, performing a 360-degree search, before proceeding on its way. At other times, one would position itself at an intersection to serve as a relay for signals from other robots.

They were going through the building room by room, using a map created earlier by three larger robots, which had scanned the offices with laser range-finders. The human commander, who'd never seen the building before the experiment, had monitored the mapping process on a computer screen. As the map took shape, he was able to direct the robots if necessary to return to a particular area for more coverage, but he didn't have to tell them how. They'd worked that out by themselves.

After a half-hour or so of searching, one of the little units discovered a suspicious object in a closet, a pink ball about eight inches in diameter. "We'd trained the vision system to recognize that color," says Regis Vincent, a project manager. "But instead of looking for something pink, it could have been designed to sense radioactive material, or heat, or a chemical device. You name it." The robot took a picture of the ball and relayed it to the human commander, as others formed a perimeter around the ball. They'd completed their mission.

"Every time we show the Centibots, people bring up *Minority Report*," Charles Ortiz, another program manager, told a reporter from the *San Jose Mercury News*. But these units were more than movie props, he says. "They represent a major contribution in distributed robotics."

Unlike the battlefield robots deployed by the military in Iraq to investigate roadside bombs, or the unmanned drones flown remotely into hostile airspace to conduct surveillance, both of which operate as single units driven by humans, the Centibots were designed to coordinate their activities as a team, responding to one another like schooling fish. Although each unit was relatively simple, with limited computational ability, as a group they were designed to tackle missions that none could handle on its own. By communicating with nearby units, for example, they could serve as a mobile sensor net to gather intelligence about the environment. If one of them malfunctioned, the others could fill in until the broken unit could be replaced. If conditions changed unexpectedly, they could reassign tasks among various units to adjust. Like a flock of starlings, they were designed to stay together, share information, and track their nearest neighbors. But they differed from natural groups in one important way: they were not fully decentralized.

In fact, in some ways, the robots were organized more like a traditional platoon of soldiers than fish. To cover a wide area quickly, the Centibots were divided into squads, each with a leader. These leaders functioned as dispatchers, distributing tasks among members of the squad and resolving any conflicts. Squad leaders, in turn, reported to zone leaders, who reported to a team leader in a clear hierarchy of authority. By organizing the robots in this way, the scientists avoided significant computational problems in their software. But in the process they sacrificed some of the flexibility of natural swarms.

By contrast, the tiny robots built by a European team of scientists were specifically designed to capture the unpredictable behavior of swarms and flocks. "Most engineers try to control robots by knowing at each instant exactly their position and everything," says Marco Dorigo of the Free University of Brussels, who coordinated groups of researchers in Switzerland, Belgium, and Italy in what was called the Swarm-bots project (we met Dorigo in Chapter 1, in connection with his earlier work on ant colony algorithms.) "We wanted to go completely in a different direction, a probabilistic approach in which robots try to do something, but if they do not succeed after some time, they start again in a different way, a little bit in the same way as ants."

The robots they built were more playful-looking than Centibots, as if they'd collaborated with a toy manufacturer. Each S-bot, as they called them, was four inches high and four and a half inches in diameter, with knobby wheels and tanklike treads that gave them the appearance of off-road hockey pucks. Red and yellow grippers on flexible arms added to the Lego-style design, which belied a wealth of high technology tucked inside, including infrared proximity sensors, accelerometers, video cameras, speakers, microphones, light sensors, and a color ring with eight LED lights for communication.

The goal was to teach these robots how to self-organize into a team, or self-assemble into a larger unit, to accomplish some task. "Let's say that a set of robots needs to reach a certain target destination, but there is a big hole along the way that they cannot traverse," Dorigo proposes. "The first robot to arrive at the hole will realize that he cannot go on. So he switches on a light that means 'I need help.' The other robots arrive, attach to him to create a bigger and bigger robot, until it is big enough to pass over the hole, and then they go over together." The combined unit was called a Swarm-bot.

Unlike the Centibots, which shared a map of the space they were searching, the S-bots functioned with only local information, the same way that fish and insects do. They also had no leaders, relying on simple algorithms to interact with one another, which sometimes gave their behavior a chaotic appearance. In one experiment with a dozen units, S-bots zipped around inside a ten-by-sixteen-foot arena in a flurry of exploration until one of them bumped into a "prey" object that had been placed in a corner. That unit then returned to a base unit, or "nest," and began to recruit other S-bots to form a chain in the direction of the prey. After enough robots were added to the chain to reach the prey, one of the S-bots grabbed on to the prey and set its color ring to red. Others joined the S-bot and, together, they eventually dragged the prey all the way to the base. In designing this experiment, the researchers had been inspired by the foraging behavior of ants, especially those that self-organize through the use of pheromones. But instead of pheromones, the robots themselves served as trail markers.

Encouraged by the success of the Swarm-bots project, which ended in 2005, Dorigo and his colleagues decided to take the concept of robotic teams a step further by creating three types of units with capabilities that mimic those of human beings. "We wanted to investigate robots that can somehow live in the human environment, and possess some of the characteristics that make a robot a humanoid, but with completely different shapes," Dorigo says. So they set out to build a flock of some sixty robots composed of what they called eye-bots, hand-bots, and foot-bots. As the name suggests, eye-bots would provide the group with sight by flying on helicopter-like rotors and attaching themselves to ceilings. Hand-bots would grab or manipulate objects, and be capable of climbing vertical structures. Foot-bots would be designed to carry hand-bots or other objects over rough terrain. Unlike the homogeneous swarm they'd created with the S-bots, these new robots represented a heterogeneous swarm that would collaborate "to solve some simple problem like finding an object placed somewhere on the shelf, grasping it, and carrying it to a target location," Dorigo says. By late summer of 2010, he hopes to have working models of these robots, which together would operate as a "Swarmanoid."

Other researchers, meanwhile, have been looking into ways that robotic swarms might learn from natural ones. At his lab in downtown Philadelphia, Vijay Kumar of the University of Pennsylvania has assembled a menagerie of mechanical beasts that crawl, climb, and fly around the room. A professor in the School of Engineering and Applied Sciences, Kumar sees a day when robots of all stripes will fill our lives, from household robots that communicate with our kitchen appliances or garage-door openers, to sentry robots that patrol our warehouses or office parking lots. During the past few years, funded largely by the military, he and his collaborators have developed software to enable groups of robots to patrol zones of interest, locate specific objects, or track intruders. One application might be to guide autonomous underwater vehicles (AUVs) for naval operations. Another might be to help a police force monitor city streets, using pilotless drones in the sky. "Imagine a 911 call that originated from Wilshire Boulevard in Los Angeles," Kumar says. "I want eyes on the target in this general area, and I want ten-second refresh rates from video cameras of every square meter from the air. I've got five drones, and those drones have to automatically configure. That's a solvable problem."

Among his collaborators recently was Stephen Pratt, a biologist from Arizona State University, who recorded for Kumar a remarkable video of ants in the desert near Tempe carrying an object back to their nest. Although the object, a four-inch-wide disk left by researchers, was small, to the ants it was bulky; it was as if a team of moving men had been asked to pick up an enormous trampoline and haul it off. The twenty or so foragers in the video distributed themselves around the edge of the disk, lifted it off the ground, and proceeded to push and pull it to the nest. Many ant species drag prey to their colony. But few lift objects and carry them in such large numbers. Of the biologists who've seen the video, none has ever seen anything else quite like it, Pratt says.

He made the video in the same desert environment where Deborah Gordon's red harvester ants live. In fact, this species, *Aphaenogaster cockerelli*, is a common neighbor of harvesters and a competitor for food, though they rarely encounter one another. Larger and more agile than harvesters, Aphaenogasters forage mainly at night, while the harvesters work the same territory during the day, protected from the heat by their coating of hydrocarbons. To gain advantage over the harvesters, the Aphaenogasters have been known to play dirty tricks, as Gordon describes in *Ants at Work*. The Aphaenogasters, she writes, "carry out the nefarious practice of blocking up the nest entrances of neighboring red harvester ant colonies at night. This means that only the late morning light reaches inside the red harvester ant nest, so the harvester ants come out later to forage." As a result, she explains, that "leaves more food for the Aphaenogaster when they come out again that night."

Competition for food probably explains the behavior of the ants in the video, Pratt speculated. "Most ants in the desert specialize in overwhelming things, cutting prey into pieces, or driving off competition," he says. "But these guys are bad at both of those things. Their strategy instead, if they find something big, is to assemble a team, pick it up, and move it. Other ants can't do that." The disk the ants were carrying, which was wiry and lightweight, was designed by graduate students in Kumar's lab in Philadelphia as a prototype for a device to measure ant strength. Pratt had coated it with sticky fig juice to make it more attractive. "Other types of ants would have just licked the goo off," he says. "But the Aphaenogasters didn't waste any time with that. They just picked it up and carried it away."

Kumar and his team wanted to figure out how the ants did that so they could build robots as agile and team-oriented. "Looking at the video, it seems that the ants find a place and grab on to it and start moving," Kumar says. "Look, this one came up and found there was no place, and so it backed off and probably tried another approach route." At one point in the video, several ants trip over an obstacle in their path and lose their grip, which causes the disk to rotate as the others keep pulling. "They're pulling like this, and suddenly the relative velocity vector changes, and then they lose their hold with their mandibles," Kumar narrates. "When they lose their grip, the thing rotates, and they're sort of scurrying to get back to where they were, but they don't. They never do."

The next step in the ongoing project would be to manufacture a device made of a polymer that would deform slightly as the ants pushed or pulled it, Kumar says. That would give his team a way to measure the deformation from video recordings of the ants as they carried it. "If you observe the deformations, you can infer what the forces are. So we can actually tell what each ant is doing." From an engineering perspective, the challenge was significant.

"If everything is known perfectly, where objects are precisely positioned, the geometry of the objects is known, the pinch points are known up front, it's a matter of programming and coordination. You can do it," says Kumar. "But everything becomes really hard when you don't know up front where the object is, what its precise shape is, where to hold on to it. This is what the ants are doing very well."

The video was a useful reminder of how far robotics still has to go before it matches the abilities of groups in nature—let alone those in science fiction movies. Although the Swarm-bots project led by Marco Dorigo had shown that a group of autonomous robots, using local information only, could coordinate their movements to accomplish a simple task, the video of the ants convinced Kumar that natural swarms still hold many secrets about decentralized behavior. "It's a classic case, I think, where the individual motion is very hard to model, but the group motion is still predictable," he observes. "As individuals, the ants may not always be doing the optimal thing. But if you average out the effect, it turns out to be the right thing."

What Are They Learning in Those Schools?

A revolution was under way in Cuba in 1954, when a Soviet scientist named Dmitri Radakov began a series of experiments in a quiet inlet on the Caribbean coast. The subjects of his experiments were "waves of agitation," but his research had nothing to do with Fidel Castro or political unrest. The agitation he had in mind took place in schools of fish startled by sound or sudden movement. When this happened, a wave of motion would spread through the group like a ripple in a pond disturbed by a pebble, but much faster, racing from one end of the school to the other in less than a second.

What impressed Radakov about such a wave was how it emerged entirely from local interactions among the fish. It began with individuals closest to the disturbance, then spread through a kind of chain reaction as other fish responded to their neighbors. Was this one of the mysterious mechanisms by which schools of fish communicated with one another and coordinated their movements? If so, he wanted to know more about it.

In his experiments, Radakov used several devices to startle groups of tiny fish called silversides (*Atherinomorus*), including a white propeller that rotated just above surface of the water. When the propeller was activated, for example, every individual within a zone of two feet or so reacted simultaneously, turning abruptly from the blades. These were the fish close enough to see the propeller with their own eyes. Beyond that, in a zone that extended as far away as ten feet, individuals changed direction as the wave of agitation reached them. These fish were picking up the alarm indirectly.

The wave traveled so fast it would have been impossible to study it with the naked eye, so Radakov filmed the experiments with a movie camera at twenty-four frames per second. Then he projected the film frame by frame onto a big sheet of paper and marked the position of each fish. In this way, he was able to measure the wave's speed at up to thirty-four miles per hour as it raced through the school—more than fifteen times faster than the maximum swimming speed of a silverside.

This phenomenon fascinated Radakov because it showed how certain kinds of information—about the approach of a predator, say—could be rapidly shared by a group, even if the water was so crowded that most of the fish couldn't see what was happening. They didn't need to see, he wrote, because the signal was transmitted short distances from fish to fish. The very structure of a school, in other words, made rapid communication possible.

"Radakov was way ahead of his time," says Iain Couzin, a biologist at Princeton University who has recently picked up where the Russian left off. "He was among the first to start thinking about collective behavior as a social medium—as a physical type of system." Tucked away in the basement of Guyot Hall on the Princeton campus is the new laboratory that Couzin and Naomi Leonard, an engineering professor at Princeton, have set up to conduct experiments with small fish like those Radakov studied. At one end of the narrow lab, a high-speed high-definition video camera hangs from a metal frame on the ceiling, pointed down on a shallow four-by-seven-foot plastic fish tank. Their plan has been to capture the same kind of information on video that Radakov had on film, but instead of projecting movies onto a wall and tracing the fish with a pencil, Couzin and Leonard hoped to get everything they needed digitally, using advanced tracking software.

In a way, what they're doing resembles what Andrea Cavagna

and his team did with starlings in Rome. But Couzin says their job is simpler, since the fish in their tank look much larger in the video camera than birds in Cavagna's still cameras. Their dark bodies also contrast sharply with the white bottom of the tank. And the fish were accustomed to swimming in only a few inches of water, which reduced their movements to basically two dimensions, greatly simplifying the mathematics of data analysis. By capturing fish behavior on high-definition video, rather than static images, they aimed to focus on the dynamics of group motion as well as group structure.

Like Radakov, Couzin wanted to know more about the rules that individual fish use to transmit information across a school. One of the main advantages of belonging to a school, after all, is that individuals pick up more signals about the environment from their neighbors than they can by themselves, whether those signals concern the location of food, the direction of a migration route, or the approach of a hungry barracuda. Since fish in a school tend to pay close attention to one another, if one of them notices something interesting, the others are likely to notice it too, which transforms the whole group into a kind of mobile sensory array. Even weak signals from the environment, if they're important, can be magnified as they race from one individual to another.

At the same time, a group is also very good at filtering out irrelevant information, biologists say. In a noisy environment like the sea, one fish might make a mistake, for example, by responding with alarm to something that wasn't really a threat. But in a large group, the chances of many fish making the same mistake are relatively small. So trivial mistakes tend to average themselves out, protecting the group from responding to bad information.

The same thing could be said about many other animal groups, from birds to caribou. When foraging in a field, for example, individual starlings in a flock spend a greater percentage of their time pecking at food than they might be able to if they were alone, because the group will alert them if any of them spots a falcon. Biologists call this "collective vigilance," and it's practiced by herding animals as well. A study of waterbucks in the Central African Republic recently showed that members of small groups spend three times as much time scanning for predators as members of larger groups. The more eyes looking for danger, the more each animal can afford to eat. (Something similar happens with groups of people, as we'll see later in this chapter. Because of our own tendency to copy others, we too are constantly picking up cues and signals about how to act, which may affect us in subtle ways, from our eating habits to our adoption of fads and fashions.)

But how do such animals work these things out? How do they find the right balance between vigilance and feeding? Does natural selection tune their behavior to benefit the group as a whole? Or is collective vigilance nothing more than a byproduct of everybody's selfish strategies? Biologists don't have all the answers yet, but researchers have come up with some intriguing theories during the past few years using computer simulations.

Not long ago, for example, Couzin created a simulation of schooling fish based on the idea that individuals were following simple rules. In a way, it resembled the model that Craig

Reynolds had created of flocking birds. Like Reynolds, Couzin imagined a kind of bubble of awareness around each individual that determined how it responded to others. But in Couzin's model there were three bubbles. The first was a "zone of repulsion" that kept each fish from colliding with its neighbors. Think of it as the individual's personal space, Couzin suggested. The second was a "zone of orientation" that pointed each fish in the same direction as its neighbors. The third was a "zone of attraction" that kept each fish from being isolated. By maintaining these three zones, Couzin's fish were able to simulate a fairly realistic school.

Something odd happened, though, when Couzin expanded or shrank these zones. When he reduced the zone of orientation to almost nothing, for example, the fish in his simulation milled around in a disorganized way, like a cloud of mosquitoes. When he increased that zone, on the other hand, they started swimming in a circular pattern called a torus. When he increased it a little more, the fish moved forward in roughly the same direction. And when he cranked it up all the way, the group shifted into the kind of tight parallel formation we associate with those shimmering schools of fish we see in nature films.

What really surprised Couzin was how abruptly the fish changed from one shape to another. At first, as he made small changes to the interaction rules, nothing seemed to have an impact. Then all of a sudden, everything changed. One moment he was watching a school imitating mosquitoes, the next it was cruising in a big circle. It was kind of a shock. "When I first saw the torus formation, I was convinced I had a bug in my program,"

Couzin says. "I spent days and days going through code." But there was no mistake. The virtual fish were just acting out a phenomenon called a phase transition.

Just as water suddenly condenses from a gas into a liquid at a certain temperature and pressure, so the fish in Couzin's model abruptly changed from a disorganized swarming pattern to a more stable torus shape. Scientists describe such a change as a phase transition because it involves a shift from one state, or phase, to another. We saw a similar shift earlier in this chapter in the flocking model created by physicist Tamás Vicsek, which was inspired by the behavior of magnetic particles. At room temperature, the particles in a typical toy magnet align themselves with one another with respect to the direction of their spin, which gives the toy its magnetism. If the magnet is heated above a certain temperature, however, the particles get knocked out of alignment by the additional energy and the toy ceases to be magnetic. The same thing happened to the school of fish in Couzin's model, when the orientation rule was decreased. The fish stopped following one another in a torus and shifted back to random swarming.

Physicists love to talk about phase transitions, because they seem almost magical compared to the everyday rules of energy and matter. In fact, there did seem to be something mysterious about the behavior of the schools in Couzin's model. "You have to remember," he says, "there was absolutely nothing in the equations of motion for this model that said, go around in circles." The virtual fish were doing it all by themselves.

For fish in the real world, a knack for such phase transitions—

changing shape in the blink of an eye—might mean the difference between being eaten or escaping unharmed. When attacked by dive-bombing puffins, for example, schools of North Atlantic herring have been known to form a kind of bubble around the birds as they rocket through the group. Other schools of fish squeeze into hourglass shapes or spray out like a fountain to confuse hungry predators. Some even resort to a dramatic move called a "flash expansion," in which the school seems to explode and then reassemble once the danger has passed. Schools perform such maneuvers, biologists say, to make it more difficult for a predator to lock onto any single target in a swirling mass of fish. And they do it entirely through adjustments in the way individuals react to one another, rather than through some form of centralized communication.

Christos Ioannou wasn't looking for extreme behavior like that when he released the first shiners into the experimental tank at the Princeton lab. Dipping his net into a stock tank, he scooped up about forty fish and gently transferred them to the larger arena, where they cautiously started exploring. After moving several more nets full, he thought the fish still looked slightly freaked-out by their new environment. But he figured they'd settle down soon. Then something must have startled the shiners, because a group began to swim in a large circle, just like the torus formation that Couzin had simulated. "We were quite surprised to see that," Ioannou says. "I don't think anyone expected it from freshwater fish in a lab in a rectangular tank. It was a nice validation of Iain's models."

What had caused the fish to do that? Ioannou figured it had

to do with their fear of the unknown. Shortly after he introduced them to the new tank, the shiners were probably paying close attention to one another to gain information about their new surroundings. That made it easy for them to flip into the torus pattern once they got startled. In the ocean you see such formations from time to time in groups of barracuda, jack, and tuna, and one reason might be strategic. "Individuals at the front of a group are typically at the most risk of encountering predators," Ioannou says. "But in a torus, there's no front. The fish can still be aligned and react to one another quickly, but no individual has to take all the risk disproportionately."

That was important to shiners, because, unlike ants in a colony or bees in a swarm, they aren't related to one another. They don't have a stake in each other's survival. "A school might look beautiful and highly coordinated as it moves through the water," Couzin says, "but frankly, if there's a predator around, you want other individuals in front of you." If an individual fish discovers that the cost of belonging to a group, such as competition for food, has outgrown the benefits of staying, it always has the option to leave, and perhaps forage on its own. Despite such selfish motives, many fish choose to stay. Their instinct to belong must be powerful, or you'd never see schools of a billion herring in the open ocean.

Protection from predators must be a main reason to school. But does being part of a school also make a fish smarter? A few years ago, Ashley Ward, a biologist at the University of Sydney; David Sumpter, a mathematical biologist at Uppsala University in Sweden; and Couzin carried out a study with groups of three-spine sticklebacks (*Gasterosteus aculeatus*), a small freshwater fish. They wanted to find out how strongly fish are influenced by one another when making decisions, and whether that helps them make better choices.

In the first set of trials, conducted at the University of Leicestershire in England, where Ward was a postdoc, he introduced groups of sticklebacks to an inch-and-a-half-long replica made of resin plastic in a rectangular tank that was divided at one end into two refuges. Once the sticklebacks got used to the fake fish, Ward slowly towed it on a line toward one refuge or the other, making it look as if the replica knew something the real fish didn't. Would any of them follow?

The answer, it turned out, depended on how many fish were in a particular group. If there were two, then they both tended to follow the replica. But as Ward increased group size to four or eight fish, only about half of the sticklebacks followed, suggesting that larger groups needed more evidence to be convinced. When a second replica was added, that seemed to do the trick, and almost all of the fish followed.

It wasn't just the persuasiveness of the replicas that made the difference, says Couzin, who with David Sumpter helped Ward analyze the data. It was the influence of other sticklebacks as well. The more fish moved in one direction, the more others felt like joining them. "You have this classic threshold response," Couzin observes. "If you're a fish in this big group and one individual goes off, you don't follow it. But if two individuals go off within a short time period of each other, then you have this cascade, and you all go."

What the fish were demonstrating, in other words, was the same kind of quorum-based decision-making that we saw ants and bees employ in previous chapters. Just as Deborah Gordon's forager ants needed to bump into a certain number of patrollers before they were ready to go outside, or Tom Seeley's honeybees needed to sense the right number of scouts at a new nest site before declaring it to be the best one available, the sticklebacks needed to see enough fish swimming off in a particular direction before they followed. Once that happened—once the stimulus exceeded their response threshold—they followed in a chain reaction of conformity. That was the key to such systems: the powerful way they harnessed positive feedback to accelera[illegible] group decision-making. "What you see here is trade-off between speed and accuracy," Couzin explains. Depending on the urgency of the situation, the fish might adjust their quorum threshold one way or the other. "That's kind of cool," Couzin finds, "because the behavior of the fish is exactly what you get in these other collective decision-making systems."

But were the sticklebacks making *good* decisions? Their reactions might be fast, but were they also smart? A second phase of experiments put their judgment to the test. In these trials, Ward added a fake predator to the mix, an eight-inch replica of a perch (*Perca fluviatilis*), which he positioned on one side or the other of the tank between the sticklebacks and the refuge. Under normal circumstances, any level-headed stickleback, whether alone or in a group, would steer clear of such a threat. But strange things can happen when individuals let peer pressure overwhelm their common sense.

When Ward towed a single fake stickleback past the fake perch, groups of two real fish didn't know what to do. Some followed and some didn't. Larger groups kept their distance. Almost none followed the replica. But when the number of replicas was bumped up to two or three, the impact, just like before, was dramatic. Now groups of two were quite happy to swim past the perch, and larger groups seemed to lose their bearings. About half stayed away, but half followed the replicas into what would normally be a perilous situation.

What really shocked Couzin was "how similar this behavior was to the previous case—how these fish would go right past the predator if only two or three others were doing so. Really, they completely discounted their own information for the social context." This was the downside, of course, of a system in which individuals took their cues mainly from one another. "When they get it wrong, they get it really badly wrong," he says.

What would happen to their decision making if they had more individuals in their groups? A larger group, Ward figured, would give the fishes a greater diversity of opinions. So many of the political and judicial institutions that people take for granted rely on similar principles of consensus decision-making, as we saw in Chapter 2. How does this relate to the way schools of fish make decisions? If groups in general are more accurate at making certain decisions, then schools of fish ought to be too.

To test this idea, Ward held a kind of beauty contest for the sticklebacks. The contestants included four sets of replicas that had either good or bad qualities. One was fat, for example, suggesting it was skillful at finding food, while another was thin. One was dark, a warning sign about its health, while another was light. One was unusually large, another medium-size. One had small black spots on its side, indicating a common parasite, while another did not. If a group was smart, Ward assumed, it would follow the good replicas and not the bad ones, even though their differences might be subtle. And if his theory was correct, bigger groups would be more accurate at this task than smaller groups, since they had a bigger pool of information to draw upon.

As he had in previous trials, Ward ran trials with groups of from one to eight members. He presented each group with an attractive replica on one side of the tank and an unattractive one on the other. After allowing the fish to get acquainted with the newcomers, Ward towed the replicas toward the refuges and watched to see what would happen. The results backed up his theory.

In almost every case, a majority of the sticklebacks preferred the good replica over the bad one. But more important, this preference became overwhelming as group size increased. The reason, Ward speculated, was that a larger group was more likely to include at least one individual who would notice the difference between the two replicas. When that fish moved in the direction of the more attractive one, other fish followed, and, once a threshold number was achieved, they all joined in on a consensus choice. Because the threshold was higher in larger groups, moreover, their decisions were more accurate.

"If you're an individual on your own, you're not very good at making such a decision," Iain Couzin says. "But as the group you belong to increases in size, your probability of making a correct choice goes up—from around 60 percent in this experiment to 80 percent or higher."

Simply by doing what comes naturally, in other words, such as swimming in the direction of a safe haven, individuals pass along useful information to their neighbors. And if enough of them do the same thing at the same time, the whole group is more likely to follow.

Or as Sumpter succinctly put it, "The group as a whole is much better at finding the fat fish."

Do What I Do

A funny video clip from the early days of *Candid Camera* was bouncing around the Web not long ago. Like many episodes of the classic TV show, it concerned a practical joke the producers had played on an unsuspecting individual in the name of catching him "in the act of being himself." Other episodes, you might recall, featured people being surprised by a talking mailbox, a door with a sign that said "Restroom" but opened into a closet, and a camera store that changed into a dry cleaner's when the customer's back was turned.

In this particular segment, called "Face the Rear," a young man wearing a jaunty hat is joined in an elevator by three men and a woman, who are all members of the show's staff posing as office workers. We don't know why the young man, who is dressed more casually than the others, is in the building or where he's going, but he appears to be slightly intimidated by his fellow passengers, who are all wearing business attire and facing forward when the segment begins.

On a cue from "Charlie," the leader of the pranksters, all four of the staff members slowly turn and face the rear of the elevator, giving no indication why. After only a slight hesitation, the young man does the same, which makes the audience laugh. He gives in to peer pressure so easily it's comical. At the same time, his reaction shouldn't be so surprising, since he's surrounded by people who appear to know what they're doing. How could he resist?

The elevator door closes. When it reopens, everyone's now facing to our left, including the young man. The audience laughs again. The *Candid Camera* crew is starting to have fun with the guy. Next they all face forward again, and the young man is right in step, even though he looks increasingly uncomfortable, with his arms folded protectively. How far can the pranksters take this?

To find out, Charlie takes off his hat, and the young man follows suit. Then, a moment later, Charlie and the two other men put their hats back on again. After stealing a quick glance at Charlie, the young man does the same. The audience cracks up. His behavior is so ridiculous it's almost touching. Every shred of pretension has been stripped away from this fellow, showing how powerful the instinct to conform can be—as natural for people, it seems, as flocking is for birds or schooling is for fish.

The idea that people tend to copy other people is nothing new, of course. Sociologists, social psychologists, political scientists, economists, and many others have studied this phenomenon from many angles. Because we're social animals, copying plays a significant role in many of the things we do, from following a crowd to the exit in a darkened theater to buying a new novel on the bestseller list. If it weren't for our tendency to copy others, we wouldn't dress according to the latest fashion, speak in slang, or rush to the supermarket before a snowstorm to buy milk and toilet paper before everyone else does.

Consider the popularity of tattoos, says Simon Levin, a biologist at Princeton. "A group of teenage girls or boys might decide they're all going to get tattoos, or wear rings, or certain clothes," he comments. "This is not an isolated incident. They do so because there are exclusive groups that form based on whether you are in or out. So there's a lot of peer pressure. Part of it's imitation and part of it's the pressure that comes from rewards that you get, or the punishments that you get if you don't follow these behaviors."

Such groups are self-organizing, adds Daniel Rubenstein, another Princeton biologist. "If you've got a positive friendship, and you feel good because you've pierced your body, or had a tattoo, with this group, and, at the same time, you're being shunned by your parents, then you start to gravitate toward the positive. If you follow a simple rule of thumb such as 'Go to where you're getting a certain rate of positive feedback, avoid areas where you're getting negative feedback,' then the herd forms around these networks."

From a practical point of view, there's often nothing wrong with following the herd, according to Duncan Watts, a sociologist at Yahoo! Research. In fact, we don't have much choice. "You can't try to solve every problem from first principles," he says. "You have to take some things for granted," and in most day-to-day situations, when you're short of information, watching what other people are doing is a reasonable strategy. "If you're driving along the highway and it's rainy and you can't see very well, and you see a bunch of red taillights in the distance, well, you slow down, because you know those people can see something that you can't see. And they're probably doing the right thing." In that sense, we often lean on the collective experience of the group, just as flocks of birds, schools of fish, and herds of caribou do. "Nine times out of ten, it's the right thing to do," Watts says, "or at least, it's a risk-minimizing strategy. Because, really, what we're all trying to do is avoid catastrophes. Anything less than that, we can sort of rationalize."

In nature, as we've seen, following the group is often a matter of life or death. A thousand pairs of eyes are more likely to spot a predator approaching than a single pair, for example. And if there's an attack, it's easier to get lost in a crowd of individuals that look the same than it is to go it alone, since the rule of the wild is that the "odd one" gets eaten. Being part of a flock, school, or herd also increases one's chances of finding food, a mate, or a migration route. Groups even develop a sort of collective memory—what Dmitri Radakov called a "fund of reflexes"—that they hold on to even after the individuals that collected the information have left (not unlike the way the citizens of Vermont have passed down a tradition of civic discourse from generation to generation, as we saw in Chapter 2).

When it comes to exploiting specific information, it takes only a few individuals in a herd (or group of people) who know something useful to steer the rest in the right direction, even if they're not trying to be leaders. This was demonstrated recently by an experiment in Cologne, Germany, with a crowd of two hundred people. At the beginning of the experiment, everyone was gathered at the center of a big circle marked with numbers like a clock. In one variation of the experiment, ten "informed" individuals were given slips of paper that said, "Go to nine o'clock, but do not leave the group," while the rest were given slips that said simply, "Stay with the group." When the researchers said "go," the group dissolved into a chaotic churning. But after a relatively short period of sorting out, the informed individuals were able to lead the others to the designated target, in the same way that the first passengers leaving a plane tend to steer the rest toward the baggage claim area, without really trying. What was noteworthy about the experiment was how few informed individuals were required: only 5 percent. "It just goes to show how information can propagate very quickly, very effectively through groups without active signaling," says biologist Iain Couzin, who helped design the experiment.

To some extent, this seems obvious. Humans are as sensitive at picking up cues from one another as any other creatures on the planet. Yet this habit is so commonplace it's easy to overlook. Recent research, for example, suggests that it plays a significant role in whether we put on weight, feel happy, or quit smoking, among other things. In a study of social networks in Framingham, Massachusetts, a small town near Boston, a physician, Nicholas Christakis, and a political scientist, James Fowler, concluded that obesity spread from one person to another as if it were contagious. By analyzing the social networks of more than twelve thousand people, they calculated that an individual's chances of becoming obese shot up by 57 percent if a friend, spouse, or relative became obese. Similarly, an individual was 15 percent more likely to be happy if he or she was directly connected to someone else who was happy, and someone's chances of quitting smoking increased by 67 percent if a husband or wife quit. What's more, these influences spread through the town's social networks like a virus, affecting not only people at one degree of separation, such as a spouse or relative, but also those at two or three degrees. "You may not know him personally, but your friend's husband's coworker can make you fat," they write in *Connected: The Surprising Power of Our Social Networks and How They Shape Our Lives*. "And your sister's friend's boyfriend can make you thin."

How is this possible? How can someone you've never met increase your appetite for fattening foods, affect your mood, or prevent you from lighting up? It's easy to see how a spouse with a taste for late-night pizza might have an impact on your waistline. But how does a friend of a friend do that if you've never shared a meal with him?

The answer, Christakis and Fowler say, is that smoking moves through a social network the way electricity moves through a power grid or humid air moves through the network of passageways in a termite nest. We don't have to see, with our own eyes, people puffing away to be affected by their habit. "What flows through the network is a norm about whether smoking is acceptable, which results in a coordinated belief and coordinated action by people who are not directly connected," they write. And the opposite is true, too. "When one person quits smoking, it has a ripple effect on his friends, his friends' friends, and his friends' friends' friends." In fact, Christakis and Fowler's study showed that clusters of connected individuals tended to quit smoking at the same time in a coordinated phenomenon that resembles the flocking of birds or the schooling of fish. People don't quit by themselves, the researchers say. It happens in cascades. "A smoker may have as much control over quitting as a bird has to stop a flock from flying in a particular direction. Decisions to quit smoking are not solely made by isolated individuals; rather, they reflect the choices made by groups of individuals connected to one another both directly and indirectly."

When it comes to more strategic behavior, however, our tendency to copy others isn't nearly as straightforward. Consider the challenges of marketing and advertising. If it's true that consumers make decisions on the basis of what others are buying, then why isn't it easier to promote products that already have a foothold in the market? For that matter, why don't consumers follow the example of influential people as often as you'd expect? Why is it so difficult to generate word-of-mouth for a product, even if it's a really good one? What role does social influence play, anyway, in determining which movies, books, or songs become hits and which become flops? What causes contagious cascades of popularity to propel certain actors, writers, or singers into superstars? And why are we so bad at predicting these things?

Intrigued by such questions, a few years ago, Duncan Watts and two colleagues set up a series of Web-based experiments using an online "music market," in which fourteen thousand people were invited to listen to songs they'd never heard before, rate them, and download their favorites. The idea was to see how strongly their preferences might be affected by those of others. Would they rate a song more highly if they knew that previous listeners had given it a good score? If so, what impact would that have upon a song's ranking or "market share"?

Subjects were recruited for the experiment from Bolt, a social networking site for young adults, while music was selected from Purevolume, a site that promoted new bands. Because they wanted listeners to rate songs that were new to them, the researchers looked for those that hadn't yet played a lot of concerts or attracted many visitors to their websites. The forty-eight they selected had names like Hydraulic Sandwich, The Calefaction, Up for Nothing, and 52metro.

As a way to measure the impact of herding behavior on song ratings, Watts and his colleagues divided subjects into two main groups: those judging songs independently and those given information about ratings by others. Since the first group had nothing to go on but personal preferences, their votes became a benchmark for quality. The second group, meanwhile, was subdivided into eight "worlds," each of which was self-contained, meaning that people in those worlds were only told about ratings made by others in the same world. This was done to compare the impact of social influence from one world to the next. If ratings were similar in all eight worlds, it would suggest that hit songs could be predicted, after all.

In the first series of experiments, subjects who logged on to the project's website, Music Lab, listened to songs from a jukebox-style menu. As each song played, they were asked to rate it from 1 for "I hate it," to 5 for "I love it." After rating the song, they were told they could download it, if they wanted to. The only difference was that subjects in the independent group were shown a menu with song titles only, while those in the social influence groups were also shown the number of times that each song had been downloaded so far. This number, the researchers figured, would serve as a social cue about the song's popularity.

After six weeks, Music Lab had collected ratings from more than seven thousand people, and the results confirmed the researchers' hypothesis—with a twist. As Watts and his colleagues had expected, in the eight worlds where each song's current ratings were made available, subjects downloaded popular songs so often they turned into runaway hits. The twist was that each world picked different songs as hits. In one world, for example, the song "Lockdown," by 52metro, was ranked number 1 by those who listened to it on the website, but in another world it ranked only number 40. Which suggested, as Watts wrote in *The New York Times Magazine*, that "if history were to be somehow rerun many times, seemingly identical universes with the same set of competitors and the same overall market tastes would quickly generate different winners: Madonna would have been popular in this world, but in some other version of history, she would be a nobody, and someone we never heard of would be in her place."

What caused this random shuffling of favorites? The answer seems to be a version of the "rich get richer" effect. As certain songs gained a slight advantage over others, for whatever reason, they benefited from that extra visibility, which expanded their lead. It didn't matter if they were good, only if they got a head start in downloads. The second phase of the experiment confirmed this hypothesis. This time, the researchers repeated the experiment but replaced the jukebox-style menu with a single list of songs ranked by popularity. Songs at the top, that is, led the rest in downloads, which the researchers predicted would amplify the bandwagon effect. In fact, this turned out to be the case.

One reason this result is unsettling is that in the real world, as we know, marketers use even stronger tools of social influence to draw attention to their products, from catchy TV commercials to product placement in popular movies, to those helpful recommendations from websites telling you what "customers who bought this item also bought." Political campaigns, meanwhile, have not only borrowed the same techniques to promote candidates, they've also harnessed the power of existing social networks to magnify the impact of their messages and enlist followers to their causes. Yet despite such efforts, predictions about which products will become hits, or which candidates will be elected, haven't gotten much better, Duncan Watts argues. "Whenever people make decisions based in part on what other people are doing, predicting the outcome is going to be susceptible to large errors, no matter how careful you are," he wrote in *The Washington Post*. "In many cases, it's probably impossible."

What does this mean, you might wonder, for professionals in fields such as investment and finance, where success or failure depends in part on anticipating the behavior of others? How do financial managers handle the uncertainties of a system in which value is established not only by the intrinsic qualities of a company or product but also by the "beauty contest" dynamics of what others think and do? Not surprisingly, one answer has been to follow the herd. If an investment manager or financial analyst wants to avoid being singled out as a poor performer, he or she will stick with the pack, even if that means blindly charging into risky situations or reflexively fleeing to safety at the first sign of trouble.

"If you're a fund manager, the name of the game is to attract new investors and not lose those you already have," says economist Peyton Young of Oxford University. "If your fund has a banner year, the word will get around and new money will flow in. But if your fund performs worse than average, current investors may pull their money out." These effects are asymmetric, he says. Typically there will be more downside risk from below-average performance than potential gains from above-average performance. Hence it's better to play it safe and run with the herd, which is why fund managers often follow similar investment strategies.

Economic forecasters, meanwhile, face a similar problem.

"You might think that, in a rational world, a forecaster would always make the best possible estimate of, say, next year's growth in GDP or next year's stock prices using the most sophisticated models and the best available data," Young says. "But that's not necessarily what they do. They might prefer to hedge their bets, because it is very damaging to your reputation when you get it wrong and others get it right, whereas it is only modestly advantageous to get it right when others are wrong. It's risky to be on the wrong side of the herd—you could be the zebra that gets eaten by the lion."

During the past few decades, as theorists have developed a greater understanding of complexity, some financial analysts have adopted more resilient strategies that consider both the pluses and minuses of herding as an investing strategy. Take the situation, not uncommon, where others have more information than you do about a particular industry or investment, says Michael Mauboussin, chief investment strategist at Legg Mason Capital Management in Baltimore. In that situation, following a knowledgeable crowd can actually be a smart move. "For instance, early investors in a promising new industry may encourage others to invest, sparking the industry's growth," he writes in his book *More Than You Know*. In other cases, following the herd can get you out of a tight spot by offering cover. "In nature, a 'follow-your-neighbor' strategy may allow a flock of birds to elude a predator. Analogously, it can help investors flee a bad investment," Maboussin observes.

For the market as a whole, however, slavish imitation is rarely a healthy thing, Mauboussin says. When investors copy one another, they automatically reduce the diversity of the market, making it more volatile and fragile. Like a nervous school of fish in uncertain waters—where every fish reacts to every little signal, whether it has relevant information or not—the market becomes twitchy and prone to sudden swings. When that happens, investors get nervous, and act out of fear rather than smarts.

Recognizing the pitfalls of herding can help investors make better decisions, Mauboussin says, because, unlike birds, fish, or caribou, people can use the gift of reasoning to counteract their instincts when it doesn't make sense. For two adventurers in northern Canada, however, the challenge was just the opposite: to see if they turn off their reasoning to tap into the inner spirit of the herd as they chased an endless sea of caribou across a frozen wilderness.

Dancing with Caribou

They'd been on the trail for a week when they spotted the first wolf. Karsten Heuer, a Canadian wildlife biologist, and his wife, Leanne Allison, a filmmaker, were skiing in the footsteps of the vast Porcupine caribou herd (*Rangifer tarandus granti*) as it migrated north to calving grounds in Alaska's Arctic National Wildlife Refuge. They had five months to go, as they followed the herd a thousand miles to the Arctic Ocean and back.

The wolf was crouching behind some rocks, but the herd had noticed it anyway. Such encounters happen from time to time during a migration, and they don't always end in an attack, so the herd was assuming a wait-and-see attitude. "There was this acceptance of coexistence, to some extent," Heuer says. "Like there are things that you fear, but they're out there all the time, and if you react to the mere sight of them, then you're going to be reacting your whole life."

As the wolf crept closer, however, it crossed a threshold in the herd's vigilance level. "Now every animal had stopped—at least the front wave of animals, the closest ones," Heuer relates. "The interior animals might still have been moving, or even feeding to some extent, but the outer ones were completely vigilant and watching." Another hundred yards, and the wolf crossed another threshold, causing the nearest caribou to turn and run. This set off a wave of motion throughout the herd until every animal was running. Here's how Heuer describes it in his book *Being Caribou*:

> It took just a fraction of a second for the closest caribou to react to the lunging wolf, and when they did, the entire mountain of animals moved in unison, erupting in a wave of flashing, spinning bodies like a turning, choreographed dance. What had been a stagnant mass of animals suddenly became a single, fleeing organism, and it veered left and right as it surged upward, pulsing silver and black like a school of darting fish.

At the rear edge of the herd, where the wolf was closing in, individual caribou suddenly split off in threads, like a blanket unraveling or tattering. "From the predator's perspective, that must have been extremely confusing," Heuer continues. "With animals heading in all directions, it must have been difficult to focus on any single individual." Chasing first one animal and then another, the wolf attempted to separate a victim from the crowd. But with each failed attempt, the wolf appeared to lose ground, until its charge dwindled to a trot. The herd raced up over the ridge, leaving the wolf panting and hungry.

In a single concise scene, this encounter between the caribou and the wolf dramatizes all three of the basic principles of adaptive mimicking. Just like the starlings in Rome, which managed to evade a falcon's attack through the *coordination* of flock movements, so the caribou responded to the wolf's charge by racing off together as if it were a single being. The initial impulse to run, moreover, was *communicated* as swiftly through the herd as waves of agitation had spread through schools of silverside fish in Radakov's experiments. Finally, by paying close attention to their nearest neighbors and *copying* their behavior, individual caribou benefit from the herd's thousands of eyes.

When the herd responded to the wolf, Heuer says, "it was like a flock of birds when they're wheeling and turning and moving, where there don't seem to be any reactions among individuals—it's all happening at once. As though every animal knows what its neighbor's going to do, and the neighbor beside that and beside that, in a fraction of a second of its happening. There's no anticipation. There's no reacting. It just is."

He and Allison noticed something else during their months on the trail. The more time they spent with the caribou, the more they began to feel as if they, too, belonged to the herd. Before they'd set out on the journey, Heuer had wondered whether this kind of thing would happen, whether, above and beyond satisfying their scientific curiosity, they'd develop an appreciation for the *intangible* benefits of being part of a herd. Heuer couldn't help wondering, in other words, what it *felt* like to be a caribou. This was a very different question than those asked by other researchers of this phenomenon. It wasn't about the mechanisms of herding, but rather the experiential side—a risky move in traditional science.

"Being trained as a biologist, I'd approached the journey very much as a scientist," Heuer comments. "We were going to go out, observe what they do, document it, and bring that story back. And what we didn't really expect, and what did happen over time, was that our identity as those scientists and documenters started to be less defined, and the boundary between us and the caribou became fuzzier and fuzzier over time." After sharing many of the same challenges with the animals, crossing the same icy streams, plodding up the same muddy slopes, suffering from the same black flies, and dodging the same bears, they felt a common bond. It was as if all their trials were "initiations for us to go through, important steps to clear our minds and open us to this level of communication among the herd, to cues and signals that we wouldn't normally process or notice."

In a practical sense, the herd functioned as an alarm system for what was happening in the neighborhood, Heuer says. "If we were with the same group for many days, especially at the calving grounds, we'd home in on the few animals on the edges of the herd that would be the most vigilant, and they would often be our cues for an incoming bear, or an incoming golden eagle, or some other threat." In an emotional sense, this translated into a feeling of increased safety among the herd. "This was nothing new in terms of animal behavior, but I was surprised that it kind of spilled over to include us as well. This may be something that biology's never going to address, but when you feel like you belong to something, it gives you so much more freedom and so much more energy that might otherwise be used up in anxiety, to do other things. For caribou, the obvious things are to feed and to rest. But for Leanne and I, it was to daydream, or to let our imaginations go a little bit farther than if we felt we were the only ones looking out for ourselves."

As Heuer admitted, he and Leanne were worn-out by the end of their grueling trek. They'd been in the wild for almost half a year, following the same centuries-old trails as the caribou. So it's no surprise, in a way, that they'd come to identify themselves so closely with the herd. But, as biologist Frank Heppner says of Edmund Selous at the beginning of this chapter, these adventurers weren't crackpots. As a trained biologist and experienced wildlife filmmaker, they knew what they were doing, and they knew what they were feeling. Whether we're ready to understand it or not, they seemed to be saying, being part of a herd can be comforting, or even nourishing, in ways that modern men and women may find hard to believe. "I think a lot of it may be biochemical," Heuer says. "The intelligence is transmitted in ways that humans don't know about or comprehend yet and maybe never will." He wasn't talking about mental telepathy, of course, or some other mystical phenomenon, but neither was he talking about flocking or herding in a strictly mechanistic sense. There was more to it, he was saying.

"This stuff is kind of mysterious, I know, and I really love that about it," Heuer says. "It's part of what keeps pulling us back toward immersing ourselves in wild places and among wild animals."

5

Locusts

The Dark Sides of Crowds

They arrived early to get a place in line. Some had camped out for days in hopes of obtaining tickets. What they wanted more than anything was to take part in a special taping of the popular game show *Wowowee*, which was celebrating its first anniversary on the air in the Philippines. Promoters had promised to give away lavish prizes during the show, including minibuses, a house, and a million-peso jackpot. By dawn on February 4, 2006, a crowd of some thirty thousand people stretched around PhilSports Arena on the outskirts of Manila, even though the stadium had room for only about seventeen thousand.

No one knows exactly what happened next, but shortly before the gates were supposed to open, the crowd became agitated. Some witnesses said a spokesman for the show had announced that only the first three hundred people with tickets would be eligible to compete for the million-peso prize. The rest would qualify for a drawing with smaller cash prizes. "They were hungry and sleepless," a public official later speculated. "When they heard this announcement, the only thing that entered their mind was that if they would not be able to enter, their hopes and dreams of getting rich would vanish."

The moment security guards opened a stadium gate—the only one being used that morning—a frantic scramble ensued. Those in the back started to push, a barricade collapsed, and the rattled guards shut the gate again. Most people in the crowd didn't know what was happening, but the throng had already been energized. Surging forward, it shoved those at the front down a steep slope toward the closed gate. Many stumbled and fell, piling up beneath the rest. The pressure was crushing. By the time it was over, seventy-four people had been trampled to death and hundreds of others injured.

Most of the fatalities, it turned out, were older women. "Their dreams must have been modest: cash to tide one over until the next day, money to pay debts off, a chance to start a small business, a house to call one's own," a newspaper editorial lamented the following day. "The roll call of their names, read over and over again on the radio yesterday—Rosalinda, Trinidad, Virginia, Nita, Elizabeth, Ofelia, Elena, Belen, Fe, Dolores—was truly dolorous, one sorrowful mystery after another."

In the aftermath of the tragedy, President Gloria Macapagal Arroyo appointed a fact-finding panel. In its report, investigators noted the stadium's lack of a written emergency plan and the inadequate number of security guards on duty for such a large crowd. The game show's host, comedian Willie Revillame, was eventually cleared of responsibility, though the network reportedly paid about 20 million pesos to help bury victims and compensate families.

Like other famous stadium disasters, the Manila stampede was a frightening reminder of the inherent dangers of large crowds. In May 2001, 130 people were crushed to death during a soccer match in Accra, Ghana. In September 2006, more than fifty people died during a political rally at a stadium in southern Yemen. There's something about a big crowd that threatens to bring out the worst in us. What had begun as a happy assembly that morning in Manila had somehow turned into a rampaging mob. Instead of making individuals in the group smarter, the dynamics of the crowd had stripped away their common sense. Instead of harnessing the power of the group to sort through bits of information and distribute problem solving, it had unleashed its chaotic energy against itself.

The same thing happens in nature sometimes when instincts go haywire. Consider the group of army ants that biologist T. C. Schneirla encountered late one afternoon outside a research facility in Panama in 1936. Schneirla, who later became a curator at the American Museum of Natural History, had noticed a swarm of small black army ants (*Lapidus praedator*, at the time called *Eciton praedator*) crossing a clearing at dusk at the Smithsonian Tropical Research Institute's lab on Barro Colorado Island, a nature reserve in the Panama Canal Zone. The next morning, the lab's cook, Rosa, came to fetch him. She wanted to show Schneirla what the ants were doing on the walk in front of the lab's library.

Several hundred of the insects were running around and around in a circle, as if caught up in an invisible whirlpool. This circle, roughly five inches in diameter, was rotating counterclockwise on the concrete, with ants near the periphery moving faster than those near the center. Occasionally, when one of the ants got turned around, it tried to move in the opposite direction, but after multiple head-on collisions with the throng, it was forced to resume its counterclockwise trek.

The army ants, it seemed, had been trapped into an endless march by their slavish instinct to follow one another. Schneirla speculated that the group may have gotten separated from its colony during a raid the day before. There had been a heavy downpour that afternoon, which may have interrupted the raid and washed away the chemical trail that connected the group with the main body of ants. When the rain had stopped, the first individuals in the group had probably set out to explore the area, sticking close to the periphery of the group, where it was safest. As they did so, they laid down a circular trail of pheromone that other ants soon followed. After a while the trail was so strong that none of them could escape.

Checking on the ants every few hours between his duties at the lab, Schneirla was fascinated by their aberrant behavior. By the end of the day, the ants had been going around and around for more than fifteen hours. When he returned the next morning, this is what he found: "On the spot of yesterday's phenomenon little or no circling is to be seen. The entire area is strewn with the bodies of dead and dying Ecitons," he wrote in his notes. "At 7:30 a.m. virtually none of the ants are left on their feet. Circling has stopped, and various small *myrmecine* and *dolichoderine* ants of the neighborhood are busy carting away the dead." The ants had followed one another to the bitter end.

This is the dark side of collective behavior. It negates everything we've seen so far about the wisdom of crowds, confirming our worst suspicions that, as one satirical slogan put it, "none of us is as dumb as all of us." When this type of dysfunction seizes a group like army ants, as Schneirla discovered, the results of collective behavior can be disastrous for every individual involved.

That's nothing, though, compared to what happens when it takes hold of a swarm of desert locusts. When these insects turn to the dark side, the damage to an entire continent can be biblical.

Dr. Jekyll and Mr. Hyde

The cloud of desert locusts that descended on the city of Nouakchott, the capital of Mauritania, in 2004 may have stretched for miles. "Within minutes the sky was brown," one resident said. "Whole trees were bending over with their weight, and the city was stripped bare of its meager greenery." People set fire to tires in the street in a fruitless attempt to drive the insects away. Children beat the air with sticks in vain. Within hours, the flower gardens at the presidential palace were all gone and the city's main soccer field was chewed to pieces.

The swarm was part of the worst plague of locusts to hit West Africa in more than fifteen years. Before the season was over, locusts would devastate crops for thousands of miles across the Sahel, reaching as far as Israel to the east and Portugal to the north, and leaving millions hungry.

"When you see locusts up close with their yellow antennae twitching, they seem possessed of an alarming intelligence," says Pascale Harter, a reporter for BBC radio who visited Nouakchott during the 2004 swarming. About four inches long, the pale pink insects looked to her like giant shrimp with wings. "Sometimes they score a direct hit, flying into your head and setting in your hair, or they crawl up your trousers, holding on with hooklike claws. And when they fill the air for hours at a time, flying in their millions in one determined direction, it feels like an alien invasion has begun."

Considering the mayhem they cause, it's easy to forget that, most of the time, the African species known as desert locusts (*Schistocerca gregaria*) are shy, inconspicuous creatures. All things being equal, these relatives of the common grasshopper prefer to keep to themselves, peacefully munching on desert plants and avoiding contact with other locusts. In this solitary phase, they represent no threat to anyone, including people. But every so often, something happens that triggers a drastic change in their behavior.

It usually starts with heavy rains. In the arid regions where they live from North Africa to India, a rainy spell can bring a sudden burst of plant life and a subsequent boom in locust populations. Eggs laid in the moist, sandy soil hatch within weeks, and a massive generation of locust nymphs appears. As long as there's plenty of food, the insects are happy. But once the desert returns to its normal, dry state and vegetation gets patchy, the locusts find themselves jostling one another for what's left. When the density level of their population reaches a tipping point, something inside of them flips.

Now, instead of avoiding one another, they actively seek out other locusts. Instead of keeping a low profile, they suddenly become more aggressive. In this outgoing, "gregarious" phase, they undergo a dramatic physical change as well. From their usual tan and green, juveniles turn a gaudy black and yellow (whereas immature adults are bright pink and become yellow once mature), signaling a radical shift from the insect equivalents of gentle Dr. Jekylls to brutish Mr. Hydes. It's one of the most remarkable magic acts in nature.

During the summer of 2004, authorities in central and eastern Australia were struggling to contain an outbreak of hundreds of small swarms of the Australian plague locust (*Chortoicetes terminifera*). "Yesterday I had a standing lucerne crop," a farmer in New South Wales told one government entomologist about his fields of hay. "Today all I have is a field full of stalks." Motorists were warned that car engines could overheat if radiator grills became clogged with the insects.

"We've got to catch them before they become adults. That's the game," says Peter Spurgin of the Australian Plague Locust Commission, which uses spotter planes to locate groups on the move, then directs crop dusters to spray them with chemical or biological pesticides. "Once they get wings and form swarms, then it's a lot more difficult, because they'll lead you on a merry chase across the countryside."

Only about a third of an inch long, juvenile locusts, called nymphs or hoppers, can form bands of seven thousand insects per square yard, many times denser than African locusts. "We can spot them from several kilometers away," Spurgin says. "Basically, they look like an oil slick on a pasture. You get this black wavy line moving through green vegetation, and behind that you have drier vegetation that has been consumed."

Scientists have known for some time that the cause of the locust's transformation from shy grasshopper to marauding plague has to do with excessive crowding. Their switch from a solitary to a gregarious state may also be a way for locusts to reduce their chances of being consumed by predators. By clumping together, individual locusts can get lost in the crowd, while, as a population, the insects can reduce the overall number of targets for hungry birds like black kites.

Until recently, however, no one knew exactly what stimulated the change. Was it the sight of other locusts? Was it their smell? Was it physical contact as they bumped into one another? To find out, an Australian biologist named Steve Simpson and several of his colleagues conducted a series of tests a few years ago to see how desert locusts reacted to various stimuli. They suspected that the key signal was chemical, but they didn't know for sure. In one experiment, they put a solitary locust in a cage where it was separated by a glass wall from ten gregarious locusts. The subject locust could see the other locusts but not smell them. It turned out that the sight of these locusts didn't have much impact on the shy locust. Next they exposed the subject locust to a chemical extract from gregarious locusts to see how it would react to their smell. Surprisingly, that didn't have much effect either. Finally, they showered the solitary insect with millet seeds over an extended period of time to simulate the experience of being constantly bumped by other insects, and that had the most impact of all, stimulating the locust to become more gregarious. It wasn't a chemical signal, it seemed, but a physical one that counted most.

Taking their inquiry further, Simpson and his collaborators sought to identify the particular body part responsible for the locust's dramatic response. This involved a tedious process of touching 170 individual locusts in specific places once a minute for four hours with a fine-tipped paintbrush. "We spent many hours tickling locusts on different body areas," Simpson says, starting with their antennae and moving backward. In each session, they tickled either the locust's face, mouth, thorax, wing pad, abdomen, foreleg, mid-leg, hind femur, hind tibia, or hind tarsus to see what might happen. The answer was not much, until they got to the locust's muscular hind leg, which turned out to be quite sensitive. After a typical session of tickling, solitary locusts touched on their thighs became significantly more sociable, preferring to join groups of gregarious locusts rather than retreat to a quiet hideaway, as they normally might. That made sense, the researchers figured, because the flank of a locust's rear leg wasn't one of the areas, like the antennae or mouth parts, that was ordinarily used for touching. If a locust repeatedly felt contact on its rear leg, it was a clear signal to the insect of overcrowding.

This signal, it turned out, triggered a release of serotonin in the locust's nervous system, a powerful brain chemical that now appears to be the key to its rapid change of personality. When Simpson's student Mike Anstey at Oxford and collaborators in the lab of Malcolm Burrows at Cambridge injected serotonin into a locust's thoracic ganglia, a cluster of nerve cells in its thorax, it induced a rapid shift of behavior toward gregariousness. When they blocked the action of serotonin in a locust's nervous system, it prevented the insect from becoming gregarious, even if it was put in a crowd or tickled.

All of which is intriguing, of course, because in the human brain, serotonin has been linked with mood, sexual desire, appetite, and some social behavior, among other things. It also contributes to the euphoric feelings of openness and sociability attributed to the "club drug" Ecstasy. Does that mean that desert locusts experience similar sensations when they change behavior? Are they on some kind of buzz when they swarm? Well, maybe, Simpson says: "Serotonin is present in many animals, and it seems to have to do with sociability, arousal, and hunger." But whether it does the same thing for locusts that it does for people is not known. What is known is that it changes locusts from individuals that detest contact with one another into ones that actively seek it out. "It transforms them from the shy individuals that sit in the corner of the kitchen at a party into the real party animals," he comments.

Once a desert locust has changed from reclusive to gregarious, it not only becomes more attracted to other locusts, it also becomes more likely to align its movement with them. And that's where the trouble can begin for people. Because if a group of hoppers senses that its location in the desert has gotten overcrowded, individuals in the group will spontaneously start moving in the same direction as the rest of the group as they seek a more hospitable environment. That can start a chain reaction, as one group stimulates another nearby group to begin hopping, eventually merging into even bigger groups, serving, in effect, as seeds for a massive swarm. Which means that the key to preventing such swarms—and avoiding vast crop losses like those suffered in Mauritania—might lie in knowing exactly what the threshold level was for locust density. When does *crowded* become *too* crowded?

That was one of the key questions Simpson was hoping would be answered when a team of his colleagues at Oxford University, led by a postdoctorate researcher named Jerome Buhl, set out to try something new. Computer models had predicted that groups of animals would suddenly change from a kind of random movement relative to one another to a highly coordinated motion as their density increased. But nobody had tested it in the lab with locusts. So Buhl and his collaborators placed groups of hoppers into a circular plastic arena measuring about three feet in diameter. These groups varied in number from five to 120 individuals. As predicted, when their density was low, the locusts tended to move around in all directions. But as soon as the number of individuals exceeded thirty—the equivalent of about sixty locusts per square yard in the desert—the locusts spontaneously started marching in unison around the circle, just as the misguided army ants had done in Panama. Sixty individuals, it seemed, was the threshold level for triggering a swarm.

But there was more to the story. One day, while reviewing a video of a routine experiment, Buhl noticed that the number of locusts in the arena was one short. He mentioned this to the researcher who'd introduced the locusts into the arena that morning.

"Well, I put them all in," his colleague said. "I put in twenty."

"Okay, but there were only nineteen when I looked at the video," Buhl said.

Being stubborn, his colleague rewound the video of the experiment to the beginning, and they counted the locusts. "Look, twenty individuals," he said. "Hah!" So Buhl fast-forwarded to the end, and they counted again. "Nineteen!" Buhl said. "What happened?"

When they watched the video together, they noticed that locusts in the arena were coming up to one another very aggressively and biting one another, sometimes as often as once every seventeen seconds. In one scene, a locust was holding down another locust and eating it. By the time it was finished, the victim was completely gone, which explained the discrepancy in the number of individuals in the video—as well as something else. If desert locusts were cannibals, as they appeared to be, that would be an excellent reason for them to be so sensitive to bumping on their rear legs. Being able to sense the approach of another locust from behind, and being able to repel an attack with a swift kick, might be a matter of life and death.

This threat of cannibalism, at the very least, cast a new light on the group's behavior. Unlike the collective efforts of ants, bees, or termites, in which self-organization springs from the instinctive commitment of each individual to the colony, locust swarming seems to be based primarily on fear. Individual locusts in a swarm aren't contributing to the greater good. They're out for themselves. They don't appear to be collaborating with one another. They're just trying to eat the locust in front of them and keep from being eaten by the one behind them.

No wonder that swarming locusts feel an urgent need to keep moving, Steve Simpson figured. They don't want to become victims in what amounts to a forced march. Locusts are so sensitive to being touched on their rear legs, he says, because "you need your back legs to kick off other locusts or you're dead meat."

Death on the Bridge

Pilgrims who take part in the annual hajj in Saudi Arabia don't expect it to be easy or comfortable. The five days of rituals in and around Mecca are opportunities for the faithful to remember the trials of Abraham and to show solidarity with their Muslim brothers and sisters from around the world, several million of whom arrive at the same time by bus or plane, or on foot. The experience is hot, crowded, physically exhausting, and emotionally draining, especially for the elderly. Yet every year, more and more people join the throngs as they move from one sacred site to another in the desert.

On January 12, 2006, several hundred thousand pilgrims had gathered in a dusty tent city at Mina, three miles east of Mecca, eager to fulfill one of the last obligations of the hajj, the ritual stoning of the three pillars called al-Jamarah. These pillars symbolize the devil, who Muslims believe tempted Abraham, urging him not to go through with the sacrifice of Ishmael that God had demanded. To get close enough to toss pebbles at the pillars, some pilgrims had to make their way up broad access ramps—in appearance not unlike those at airport terminals—onto a concrete platform that surrounded the pillars. Two years earlier, 249 people had been crushed to death by a crowd on this span, which people call the Jamarat Bridge. Some officials blamed peddlers selling food and souvenirs for clogging up traffic. Others pointed to pilgrims carrying personal belongings. Some people complained about the lack of signs telling pilgrims where to go, while still others said the crowd ignored the few signs there were, because too many people spoke different languages. The situation didn't appear to be completely under control.

Anticipating another massive crowd in 2006, Saudi authorities had made several changes. They'd enclosed the traditional narrow pillars with elliptical walls to improve the flow of traffic and give a greater number of pilgrims access to them at the same time. They'd installed a system of video cameras to monitor crowd movements, put up barriers to direct traffic, and assigned some ten thousand troops to maintain order. As soon as the hajj was over, they planned to tear down the thirty-two-year-old bridge and replace it with a larger, multitiered structure that would vastly improve the flow of the crowd. But for now, they'd done what they could. Would it be enough?

By noon on January 12, about a half-million or more pilgrims filled the Jamarat plaza in front of the bridge, men wearing simple long white robes, women in traditional garb. In the past, some sects had issued religious edicts forbidding the stoning ritual until after the midday call to prayers, but this year some pilgrims had gone ahead that morning. Shortly after noon, the main ramp leading onto the bridge from the east was packed solid with people.

About that time, something happened on the bridge that caused people to jam up. The first reports were confusing. One Saudi official said the bottleneck was caused by pilgrims tripping over bags again. Others claimed that security forces, for some reason, had blocked the entrance. Whatever the reason, the holdup quickly became serious, because the enormous crowd kept pushing forward.

"I saw people moving and suddenly I heard crying, shouting, wailing. I looked around and people were piling on each other," an eyewitness told the BBC. People were shoving, elbows flying. Within minutes, ambulances and police cars were racing to the scene. But the pressure inside the crowd was crushing, making it difficult for safety personnel to get to those in trouble. By the time they did, more than an hour later, victims were piled up in layers seven deep: 363 men and women were dead, including several security guards. Hundreds more were injured. "We tried to encircle the women to protect them," an Egyptian pilgrim later told a TV station from his hospital bed. "But it was useless. They are all gone."

The grieving families that came to the morgue outside Mina to claim loved ones had traveled to the hajj from Egypt, Pakistan, Bangladesh, Yemen, Sudan, Afghanistan, and several other countries, as well as from Saudi Arabia, reflecting the indiscriminate toll of the accident. Saudi security forces had done what they could to prevent this very thing from happening. Yet it had happened anyway. "It was like the road of death in there," one pilgrim told Reuters. What had gone wrong?

In search of an answer, Saudi officials asked a team of European experts to analyze surveillance video from the accident. By chance, a camera on a 115-foot pole directly above the ramp had recorded the whole thing. One of the researchers enlisted was Dirk Helbing, then of Dresden University of Technology, a physicist turned sociologist, who was doing traffic science and crowd research at the time. Unlike other analysts, who have approached the subject from a psychological, sociological, or engineering perspective, Helbing has investigated parallels between crowd movements and the behavior of particles or fluids. In his earlier research, he'd applied formulas from fields such as molecular dynamics, kinetic gas theory, fluid dynamics, or granular flows to create computer simulations of the way pedestrians spontaneously form lanes to get by one another in crowded crosswalks or the way that traffic jams suddenly appear on open highways for no apparent reason. Such research had persuaded him that it can be useful, under certain circumstances, to think of a crowd as a gas, a liquid, or a solid.

"It depends on the density of the crowd," he says. "Let's say we're studying pedestrians entering a stadium. At low densities, people can move freely without a lot of interactions with each other, so that's a kind of gaseous state. When the density becomes higher, people obstruct each other and there are mutual interactions, so one could call this a fluid state. At even higher densities, particularly if people are pushing each other, then you can use the granular analogy to understand what is going on." Like grains of sand in an hourglass, in other words, people trying to move through a bottleneck can clog it up so tightly that no one can move.

Helbing expected to see such patterns in the video of the accident. But the results of the analysis didn't turn out exactly the way he'd anticipated. The footage that he and his colleagues analyzed showed the right half of the ramp leading onto the bridge. At 11:45 on the morning of the accident, a throng of pilgrims was moving along relatively smoothly on their way to the pillars. There was even space for some individuals to carry umbrellas to shade themselves from the sun. Then, at 11:53, those in front suddenly came to a halt, possibly because a first-aid and help center set up at a corner was in their way. They didn't stay like that for long, resuming their forward motion in less than a minute. Yet it was long enough to start a sequence of stop-and-go waves that flowed upstream behind them. This behavior, in a way, looked similar to what Helbing had observed in his studies of highway traffic. When a driver sees brake lights ahead, he taps his own brakes, which prompts drivers behind him to slow down as well. But this delay, rather than being caused by speed adjustments, may have been created by people competing for the few gaps between them.

The stop-and-go pattern on the ramp continued for more than twenty minutes, as wave after wave rippled through the crowd, and density levels continued to build. (To measure density and other factors, Helbing and his team had analyzed video recordings frame by frame using tracking algorithms that recorded the number of people entering the bridge, their speeds, flow rates, and local densities.) Then suddenly everything changed again at 12:19, as the crowd shifted into what Helbing described as an "irregular" flow pattern. In this phase, which he compared to turbulence in a fluid or an earthquake, pilgrims were jammed up against one another, and clusters of people got randomly shoved this way or that against their will. In such a situation, as another expert once described it, "shock waves can be propagated through the mass sufficient to lift people off of their feet and propel them distances of three meters [ten feet] or more. People may be literally lifted out of their shoes, and have clothing torn off."

Within minutes, several pilgrims had stumbled and fallen. Unable to get back on their feet, they were trampled as the enormous force of the crowd surged forward. More individuals fell, becoming obstacles for yet more pilgrims, and, in this way, the disaster snowballed. It was a vicious cycle: the more the crowd jammed up, the more it reduced the ramp's capacity to move people onto the bridge, the more compression levels increased, until the crowd's density reached "criticality" and people started to die. Even those who didn't fall were in mortal danger of being crushed by the pressure of the throng, which in similar circumstances has been powerful enough to bend metal or knock down a brick wall. "The impact of the forces can best be compared with a car resting on your chest," Helbing says. "But its direction is changing all the time."

What surprised Helbing most about the video was how long the warning signs persisted before the accident took place. The stop-and-go waves on the bridge ramp lasted for more than twenty minutes. And the turbulent pattern that followed went on for another ten minutes. If the Saudis had used a video analysis system that showed the crowd's flow rate in real time, he speculated, security personnel might have had thirty minutes in which to do something. They could have reduced the flow of pilgrims entering the bridge, for example, or rerouted some of them, or separated them into blocks to keep shock waves from propagating. They had enough time.

Maybe not, cautioned another crowd expert who has also consulted for the Saudis. "By the time you see turbulent shock waves like that, it's too late," observes Keith Still, founder of a British firm, Crowd Dynamics, that had helped the Saudis design a new traffic plan for the stoning ritual a few years ago. "What are you going to do with a crowd that's in this extreme shock-wave state? It's already at the point of collapse. Anything you do may well be the thing that causes that final push to create the accident."

Those in charge need to remember the root cause of the problem, Still says: too many people trying to get through too small a space. "The ingress rate at the bridge was 135,000 per hour. The throughput rate of the pillars was only 100,000 an hour. You can't pour a pint into a half-pint jug."

The Saudis knew the bridge was outmoded, according to Still. As soon as the last pilgrim left the span in 2006, they were going to demolish it and begin work on a new one. That was why they'd cleared a larger space around the entrance ramp to make room for construction equipment. Unfortunately, by doing so, they'd also made room for a larger crowd to gather in the meantime. By the time the stoning ritual began, there may have been 750,000 in that plaza, he estimated. Combine a crowd of that size with a religious edict telling pilgrims to wait until the midday call to prayers to begin the ritual, and you have all the makings of a disaster. It was like launching the biggest marathon race ever held. "Bang goes the gun, up goes the call for prayers, everybody tries to move forward, and the design just can't cope with that," Still says.

Under normal circumstances, a crowd can handle a surprising amount of congestion through self-organization. When two streams of pedestrians meet, for example, they naturally sort themselves into separate lanes, one person following another, to reduce obstructive interactions, just as army ants do in the rain forest. But when density greatly exceeds capacity, even temporarily, the dynamics of a crowd changes, and individuals lose control, both physical and psychological. At extreme densities, individual motion and awareness is replaced by mass motion and response. The crowd loses some of its ability to self-organize and may become an unthinking mass, driven by physics and fear. Instead of breaking down traffic problems through simple rules of thumb and distributing them among countless individuals, as an ant colony does, the crowd flips into a chaotic state driven by self-preservation—more like a swarm of locusts.

One reason for this breakdown is a lack of communication among those in the crowd. In an assembly so large, information doesn't propagate from one person to another as rapidly as it does through a school of fish or a flock of birds, animals whose instincts have evolved to pick up such signals (as we saw in Chapter 4). In particular, critical information doesn't spread from the front of a crowd to the back, which makes it impossible for security staff at the scene of a trampling to reduce pressure, because those doing the pushing are out of hearing range. What often happens instead is that, in the absence of reliable information, rumors can form and spread, raising anxiety levels so high that people act irrationally and start pushing, shoving, or trying to run away.

The only sure way to prevent such chaos at mass gatherings is to design and build in safety factors from the start, which is what the Saudis set out to do immediately after the accident in 2006. The old Jamarat Bridge was taken down and a new one started that would eventually have five levels or more instead of the previous two (the ground level and single elevated span). Besides providing more capacity in general, each level would have multiple entrances and exits. New traffic strategies were also put in place, including the installation of fences to channel pilgrims directly from the vast tent city to the bridge, and the rerouting of all pathways to make them one-way. Registered groups of pilgrims would be assigned separate starting times through a new scheduling system, and these times would be spread out through the day. Security personnel were trained to use surveillance systems that automatically counted pilgrims. "It wasn't just the bridge, it was an entire approach," says Keith Still. "Everything was now under complete control."

Maybe so, but by the time the next hajj was held in December 2006, only the first two levels of the new bridge had been completed, and the crowd that assembled was one of the biggest ever. More than two million people jammed into the valley around Mina this time, giving the new Jamarat Bridge its most severe test yet. Anxiety levels must have been high for pilgrims, who remembered the hundreds just like them who had lost their lives less than a year before. Nevertheless, the stoning ritual went forward.

But this time there was no accident. No trampling. No panic. Although the crowd was as vast as any authorities had ever seen in Mina, its density levels never exceeded the capacities of the new structure—as the design intended. And even though people in the sea of pilgrims were hot and exhausted, they never lost their sense of identity. They never became a mindless throng. They did what they had come to do and then they went home, tired but happy.

The Saucepan Revolution

The crowd outside the parliament building in Reykjavík was angry but well behaved. Bundled up against the cold, they banged pots and pans with wooden spoons, blew whistles, shouted slogans, and tossed snowballs at the gray stone edifice where Iceland's legislators were gathering for the first time in 2009 since their winter break. "We have to wake up the government that seems to be fast asleep," a woman told a reporter from *The Sunday Times*.

A few months earlier, Iceland's high-flying banking system had crashed after getting caught in a variety of risky borrowing schemes. Government regulators had been forced to seize the country's biggest banks, which together had piled up debts totaling many times the nation's cash reserves. As foreign investors fled, Iceland's stock market had plunged by 85 percent and its currency, the krona, had lost a third of its value in a single week. For average citizens like those in the crowd, this financial meltdown meant widespread layoffs, vanishing savings, plummeting real estate values, impossible mortgages, and sky-high inflation. No wonder they were steamed.

"People are simply so frustrated," said another protester, who was slamming two pot lids together. "A whole nation capsized, and everyone from the government to the central bank and the financial supervisory authority are still in their positions."

Not for long, as it turned out. A few days after the protest—

part of a movement that came to be known as the Saucepan Revolution—the prime minister, Geir Haarde, announced that he and his whole cabinet were resigning. Theirs was the first government to be brought down as a direct result of the global financial crisis. Iceland had gone from being one of the world's most prosperous nations to a financial basket case in a matter of months.

This sudden collapse was a failure of market behavior as dramatic as the worst stadium disaster was of crowd control. When a market functions properly, experts tell us, it balances fear and greed automatically, sorting out the competing desires of a multitude of investors through their separate calculations of profit and risk. But when it gets out of balance, when investors blindly follow one another—either in hopes of riding a bubble or in panic when the bubble bursts—a friendly marketplace can turn into a roaring stampede.

The story of how this happened in Iceland goes back to the late 1990s, when the nation deregulated, then privatized its banks. Buoyed by high interest rates fixed by the government to manage inflation, the banks suddenly became aggressive players in international finance, offering irresistible deals to foreign depositors. In less than five years, Iceland's three biggest banks pumped up their combined assets tenfold in one of the fastest expansions ever seen in global banking. The same high interest rates also helped strengthen the krona, which prompted Icelanders of all stripes to go on spending sprees or take out loans and mortgages in foreign currencies. Awash in money from abroad, the island's 300,000 or so people, who for centuries had earned their living mainly from fishing, started to believe that Iceland's financial bubble would never burst.

Then, in October 2008, it did.

It began with the sudden tightening of credit worldwide. As the mortgage-securities crisis spread from Wall Street across the globe, banks quit lending to other banks, catching Iceland's Glitnir bank by surprise. Unable to pay back sizable short-term loans, the bank agreed to be nationalized. That panicked customers, who made a run on deposits not only at Glitnir but also at Iceland's other two largest banks, Landsbanki and Kaupthing. When they too went under, the government guaranteed domestic deposits, but not those in the banks' subsidiaries overseas. Foreign depositors were out of luck, including 120 British municipalities that reportedly lost $1.8 billion in savings. In retaliation, British authorities seized the assets of a subsidiary of Kaupthing bank under an antiterrorism law.

One irony of this mess was that it had nothing to do with the mortgage meltdown that was causing havoc across the rest of the world—at least not directly. Despite the reckless abandon with which Icelandic bankers had accumulated foreign debt, they'd avoided, for the most part, the American subprime mortgage schemes that had proved to be so toxic. In that sense, Glitnir and the other banks were accidental casualties of a cascade of interconnected global events that came tumbling down on them like a string of giant dominoes. The bankers' mistake was not recognizing, long before that happened, that their nation's financial bubble was unsustainable. "Who knew that Iceland was just a hedge fund with glaciers?" wrote *New York Times* columnist Thomas Friedman.

Going on television to explain the crisis, Prime Minister Haarde told fellow Icelanders that seizing the banks had been necessary to prevent the nation's economy from being "sucked with the banks into the whirlpool." The lesson to be learned, he said, was that "it is not wise for a small country to try to take a leading role in international banking."

Maybe so, but there was another lesson here: one that reminds us of the enduring appeal of get-rich-quick schemes. By leveraging massive amounts of debts off the sudden influx of deposits, Iceland's bankers, dubbed the New Vikings, had joined a club of big-time speculators going back centuries. In 1720, for example, fast-talking money men in England raised an astonishing amount of cash from investors by securing from the crown the exclusive rights to trade with Central and South America—even though that part of the world was under the control of a hostile Spain. They were so persuasive selling the mere idea of the deal that stock in the company shot up tenfold in just a few months, which encouraged others to give similar ventures a try, until almost everyone in England was risking their savings on one dodgy scheme or another. By the time the "South Sea Bubble," as it was called, finally burst a few months later, the scandal had spread from financiers to government officials, who were impeached for their complicity.

It was even worse during the tulip craze that seized Holland during the early seventeenth century. In one of the most celebrated speculative bubbles of all time, traders bid up the price of the bulbs until they reached a hundred times their weight in gold. Although tulips had been cultivated in Persia since the eleventh century, and in the gardens of Suleiman the Great in Istanbul since the early sixteenth century, the Dutch had never seen flowers so striking, as Mike Dash writes in *Tulipomania: The Story of the World's Most Coveted Flower & the Extraordinary Passions It Aroused.* "The colors they exhibited were more intense and more concentrated than those of ordinary plants; mere red became bright scarlet, and dull purple a bewitching shade of almost black." For three decades at the beginning of the seventeenth century, wealthy collectors vied for the rarest specimens. This was a time of great prosperity in the Netherlands as Dutch ships brought back fortunes in spices and exotic goods from Indonesia and other parts of Asia. Cash flooded through the economy. Then in 1635 speculators got into the game.

The dangerous innovation they brought was that, instead of buying and selling actual bulbs or flowers that physically changed hands, they started trading in bulbs that were still in the ground. Often, they were allowed to put down a small deposit, with the balance due months later when the bulbs were lifted from the soil. When people realized how easy it would be to make a quick profit this way, everybody wanted a piece of the action, from carpenters to bricklayers to lawyers to clergymen. Weavers sold their looms to make down payments, assuming that by the time the tulip bulbs were raised they could sell them for much more. This was the same kind of leveraged gamble, of course, that got Icelandic banks in such trouble, and it led Dutch speculators into deep water too. Before long, the price of tulips bore no relation to the actual flowers. "It became perfectly normal," Dash writes, "for florists to sell tulips they could not deliver to buyers who did not have the cash to pay for them and who had no desire ever to plant them."

By 1636, the price of the most desired tulips had reached ludicrous heights. One pamphleteer at the time noted the sale of a single bulb for three thousand guilders, listing the equivalent in goods that could be purchased for that amount: eight fat pigs, four fat oxen, twelve fat sheep, twenty-four tons of wheat, forty-eight tons of rye, two hogsheads of wine, four barrels of beer, two tons of butter, a thousand pounds of cheese, a silver drinking cup, a pack of clothes, a bed with mattress and bedding, and—last but not least—a boat.

When the bubble popped in 1637, it happened almost as quickly as the financial meltdown in Iceland. In a few shocking weeks in February, tulips that had each been worth twenty times a carpenter's annual wages were suddenly worthless. Panicking traders tried to sell their overpriced bulbs at any price. It was as if, sensing that prices had gotten out of control, Dash writes, "tulip traders had been waiting for something to happen, and now it had." The same thing had occurred during the worldwide banking credit crunch of 2008: suddenly, nobody wanted anything to do with risk. Then, as now, if you were leveraged with debt, you were out of luck. When the dust finally settled in Amsterdam and collectors returned to the market to pick up the pieces, most bulbs could be had for 5 percent if lucky and often for 1 percent or less of their former values.

It was yet another chapter in "the great and awful book of human folly," as Charles Mackay writes in his celebrated 1841 account, *Memoirs of Extraordinary Popular Delusions*. "Money, again, has often been a cause of delusion of multitudes. Sober nations have all at once become desperate gamblers, and risked almost their existence upon the turn of a piece of paper." What makes such bubbles so dangerous is the same thing that makes a big crowd so dangerous: you never know when a small event, like a jump in foreclosure rates in California or a pilgrim's bag left on a pedestrian ramp, will trigger a sudden, crushing collapse.

"Bubbles and crashes are textbook examples of collective decision making gone wrong," James Surowiecki explains. "In a bubble, all of the conditions that make groups intelligent—independence, diversity, private judgment—disappear." Instead, individuals take their cues entirely from others, blindly chasing a quick profit like the ants that went around and around in the circular mill that T. C. Schneirla saw in Panama. "It's already hard enough, as we've seen, for investors to be independent of each other," Surowiecki writes. "During a bubble, it becomes practically impossible. A market, in other words, turns into a mob."

Something similar happens with locusts. "When individual insects run short of resources, they try to cannibalize each other," Iain Couzin says. "This leads to positive feedback and highly destructive behavior. When a mob gets the same kind of positive feedback it also leads to aggressive, violent, or destructive behavior." In both cases, something small can grow rapidly into something huge. Just as a crowd disaster scales up quickly to create massive casualties, locusts scale up quickly to create massive swarms.

The key to preventing crowd disasters, experts say, is to make sure that densities stay below critical levels. This can be accomplished, Keith Still advises, by intelligently designing stadiums and other structures where the crowds gather to handle maximum capacities. A crowd can also be defused by providing enough information to individuals to maintain personal control. "The crowd itself doesn't realize the dangers or the risks because they have limited sight, limited ability to move," Still says. When you're in the middle of the pack, "you're constrained by the crowd. You're constrained by the fact that you can't take a whole pace forward." Under such circumstances, if you can't get accurate information about what to expect, you can lose your sense of control and go with the flow in the mob.

Information is crucial to healthy markets, too. When investors give up personal judgment in favor of following a crowd's whim, they disconnect a market from critical information. "The problem is that once everyone starts piggybacking on the wisdom of the group, then no one is doing anything to add to the wisdom of the group," Surowiecki writes. In the most extreme cases, a giant bubble can form, taking an entire country on a wild and bumpy ride.

In the general election that was held in April 2009, Iceland's voters booted out the conservative party that had controlled parliament for nearly two decades, giving leftists thirty-four of sixty-three seats in the body. Promising to pay back $5 billion in loans from the International Monetary Fund and Nordic neighbors as quickly as possible, Prime Minister Johanna Sigurdardottir pledged to seek membership for Iceland in the European Union. With any luck, she told reporters, "the first country to have an economic crisis on this scale" would also be the first to recover.

"We have grown used over our history to bad harvests, seasons with no fish, the bad climate, things going up and down," Olafur Hardarson, dean of social sciences at the University of Iceland, told *The New York Times*. "People are saying, 'This will be bloody tough, but we've got to get on with it, and we'll muddle through.'"

Conclusion

Doing the Right Thing

On Memorial Day in 2009, Yale University conferred the honorary degree of doctor of social science upon the Nobel Prize–winning economist Thomas Schelling. A former Yale professor, Schelling was one of ten people recognized that morning as part of the school's commencement ceremony. Secretary of State Hillary Rodham Clinton was another.

Schelling didn't know it as the event got under way, but he was about to take part in the kind of intriguing social situation he's famous for analyzing—one that takes place in the most familiar of circumstances but raises deeper questions about an individual's role in a world of group behavior.

"As each of us was called to walk up and shake hands with Yale's president and receive a diploma, there was applause," Schelling says. "They had saved Hillary Clinton for last. When her name was called and she walked up to where the president stood, the audience not only applauded but a whole bunch of students—there were thousands in this huge open square—began to stand up. And pretty soon more students stood up. And, as it caught on, almost everybody stood up."

Behind Schelling on the stage, where he'd been seated with other dignitaries and school officials, a few people also rose to their feet. Should he stand up, too? "For a lot of us, it was a puzzling choice," he says. "If only three of us on the stage stand up, what does that say about us? What does that communicate to the ones that didn't stand up? Were we showing off, or what?" By now, the question had nothing to do with whether the secretary of state deserved a standing ovation. It was all about group dynamics.

Like many of the phenomena we've seen in smart swarms, a standing ovation depends on a series of interactions among a number of individuals, as economists John Miller and Scott Page explain in their book *Complex Adaptive Systems: An Introduction to Computational Models of Social Life*. First, one or more members of the audience must feel strongly enough to break the ice and be the first to rise from their seats. Second, these people must be located where others can see them (standing ovations are rarely sparked by those in the last row). Third, a certain percentage of the rest of the audience must be in the mood to stand, once someone else gets it started. And fourth, the rest of the audience must feel enough pressure from those standing that they'd rather join them than draw attention to themselves by staying seated.

"Though ostensibly simple, the social dynamics responsible for a standing ovation are complex," they write.

The ovation for Hillary Clinton, for example, started in the section reserved for law students. They probably leaped to their feet at the announcement of her name because she had been one of them, earning her law degree at Yale in 1973. At almost the same moment, people seated behind Clinton on the stage, presumably admirers of her career, also rose, signaling to those in the audience that a standing ovation was under way. "This was a complicated one," says Schelling, "because there was not only group behavior, but different groups—people on the stage as well as people in the audience. Pretty soon I looked around and everybody was standing up, and so I stood up in order not to be conspicuously sitting down."

This was precisely the type of situation Schelling has analyzed so brilliantly, in which individuals make decisions based not only on their own feelings and opinions but also on the actions and reactions of others. As a pioneer in applying game theory to behavioral economics, Schelling has written about intractable issues from nuclear weapons to climate change. Yet even he was momentarily baffled by the way things unfolded. If Clinton had been the only person being honored that day, a standing ovation would have come as no surprise. Considering her celebrity, decades of public service, and position in the government, no one could doubt it was the right thing to do. What had thrown Schelling off was the fact that nine previous honorees, all extraordinary in their own rights, had received the same recognition without standing ovations, until the law students, primed for action, forced the crowd to reconsider that morning's rules of behavior.

Such situations are worth talking about, Miller and Page say, not because we care so much about standing ovations, which are fun to take part in and even more fun to receive. They're worth talking about because the same pattern of behavior that causes them—individuals influencing one another through a cascade of signals or actions—also has something to do with whether people send their children to public schools or private schools, cheat on their taxes, keep their dog on a leash, put on weight, pass along a rumor, vote for a particular party, or decorate their houses with Christmas lights. To understand the workings of a standing ovation is to better understand how Twitter becomes a household name or a local outbreak of swine flu becomes a global pandemic. "Although standing ovations per se are not the most pressing of social problems," they write, "they are related to a large class of important behaviors that is tied to social contagion."

A standing ovation, that is, demonstrates in a highly simplified way the mechanisms by which contagious behavior works. In that sense, it plays the same role that flocks, herds, and colonies have played throughout this book as models of different kinds of dynamic group behavior. By observing ant colonies, for example, we've seen how a large number of individuals without supervision can accomplish difficult tasks by following simple rules when they meet and interact. Because a colony distributes problem solving among many individuals, it not only allocates its resources efficiently but also adjusts rapidly to changes in the environment. It was the clever way that ant colonies instinctively self-organize, after all, that inspired computer scientists to capture their behavior in algorithms, which ultimately helped a company like Air Liquide optimize its complex business operations.

From honeybee swarms we've learned that groups can reliably make good decisions in a timely fashion as long as they seek a diversity of knowledge and perspectives, encourage a friendly competition of ideas, and narrow their choices through a mechanism like voting. This process of deliberation not only suggests a practical way to tap into the wisdom of crowds at a big organization like Boeing, as Dennis O'Donoghue recognized, it also helps communities build up intangible reserves of trust that they can draw upon in tough times, as the citizens of Vermont discovered.

By studying termite mounds we've seen how even small contributions to a shared project can create something useful and impressive when large numbers of individuals build upon one another's efforts. In the same way, Web-based tools such as wikis, blogs, and social tagging have offered intelligence analysts effective platforms for pooling information and improving on one another's insights.

Finally, flocks of starlings have shown us how, without direction from a single leader, members of a group can coordinate their behavior with amazing precision simply by paying close attention to their nearest neighbors. Through the same kind of adaptive mimicking, a school of fish or herd of caribou can rapidly communicate information across a group, whether it concerns the location of a food source or a migration trail or the approach of a predator. These last groups remind us that we're social animals too, constantly picking up cues from one another about how we should behave, whether we should cut into a long line waiting for tickets, reduce our carbon footprint, or give up cigarettes for good. Paying close attention to others can be a handy decision-making shortcut during moments of uncertainty, we've learned, but it can also tempt us to follow the crowd uncritically, such as when we get swept up in fads or reckless financial schemes.

Compared with the clever things that natural swarms do, standing ovations look haphazard and arbitrary—like many other situations in which our behavior is strongly influenced by what other people are saying and doing. More often than you might expect, as computer simulations have demonstrated, our tendency to mimic one another causes ovations to misfire. People end up standing and clapping for performances that few have actually enjoyed, and the reason might be as simple as the fact that friends of the actors have managed to get a cascade started. "On the other hand," says John Miller, "you could get an audience that loved the performance who all stay seated, because they're in Minnesota or something, and they're too shy to stand up." Groups taking part in a standing ovation, that is, behave quite differently from those tapping into the wisdom of crowds. Instead of combining diverse bits of information and perspectives into a single smart decision, audiences in an ovation can be pushed one way or the other by waves of mimicking. When that happens, the results can be unpredictable, sometimes expressing the exact opposite of how the group really feels.

That would never happen with a swarm of honeybees, biologist Tom Seeley says. "If worker bees formed the audience at the commencement ceremony at Yale, each one would have taken note at the start of the standing-up process that some members of the 'hive' were changing posture and therefore that something was up, but each bee would not have felt obliged to follow suit," he explains. They just aren't genetically programmed to do that. "Instead, worker bees have been shaped by natural selection to avoid such peer-pressure/hop-on-the-bandwagon effects, and so to make their individual decisions independently." When a scout learns about a potential nest site by watching the dancing of another bee, as we saw in Seeley and Visscher's house-hunting experiments, she indicates support for the site only if she has evaluated it for herself and judged it to be desirable. In that way, Seeley says, the bees effectively combine working together and working apart. "So I suspect that the bees would have given a split decision about Hillary, with some standing with wings whirring in applause and others seated quietly, perhaps grooming their antennae."

For better or for worse, humans don't behave the same way. To vastly oversimplify our dilemma, we're torn between belonging to a community and maximizing our personal welfare. We need more than our natural instincts to help us work toward common goals. We need things like legal contracts, taxes, laws against littering, and social norms about waiting your turn and not talking during the movie. "A good part of social organization—of what we call *society*—consists of institutional arrangements to overcome these divergences between perceived individual interest and some larger collective bargain," Schelling writes. In his book *Micromotives and Macrobehavior*, he describes driving back from Cape Cod one Sunday afternoon and getting caught in a mile-long backup caused by a mattress lying in the middle of the road, apparently having fallen off another vacationer's vehicle. As hundreds of cars waited in line, one driver after another carefully swerved around the mattress, resumed speed, then drove away. All it would have taken to eliminate the backup would have been for one of those drivers to stop and drag the mattress out of the way. But there was no reason for anyone to do that once they'd gotten past the mattress and the road ahead of them was clear. So the backup continued.

In such situations, Schelling tells us, people need help to do the right thing. We need laws, norms, or financial incentives to correct the imbalance between personal and group interests. If there had been a way for drivers to reward someone for removing an obstacle from a roadway, perhaps it would have been taken care of right away, and Schelling and his fellow vacationers wouldn't have had to wait in the hot sun. But in the absence of that, the system stumbled.

Unlike ants, bees, and birds, we often don't know the right thing to do. Caught up in our own complex systems, we struggle with a lack of information, a poor sense of how one thing affects another, and an inability to predict outcomes, whether we happen to be students in Boston trying to identify terrorists, test pilots and engineers in Seattle playing the Beer Game, control-room operators in Akron struggling to gain control of a failing power grid, undergraduates in Philadelphia puzzling over network games, protesters in Reykjavík facing a banking system that's gone belly-up, pilgrims in Mina pushing across a dangerously crowded bridge, or guests at a graduation in New Haven wondering whether to stand and applaud Hillary Clinton. For all of us, sooner or later, doing the right thing is anything but simple.

In this respect, swarms in nature have taught us two lessons. The first is that, by working together in smart groups, we too can lessen the impact of uncertainty, complexity, and change, whether our groups are small ones like the problem-solving teams at Boeing or huge ones like the multitudes who maintain Wikipedia. Much depends on what we're trying to accomplish and how we structure our groups. As biologists have shown us, flocks, schools, and colonies derive their resilience and flexibility primarily from the mechanisms they use to manage interactions among individuals—what Tom Seeley calls a colony's bag of tricks. These mechanisms vary widely, depending on the particular problems swarms are dealing with, but they often include a reliance on local knowledge (which maintains a diversity of information); the application of simple rules of thumb (which minimizes computational needs); repeated interactions among group members (helping to amplify faint but important signals and speed up decision making); the use of quorum thresholds (to improve the accuracy of decisions); and a healthy dose of randomness in individual behavior (to keep a group from getting stuck in problem-solving ruts). By carefully applying similar principles in their own organizations, as Jeff Severts discovered at Best Buy, businesses can tap into the wisdom of the crowd in their own ranks, or, as CIA analysts learned, provide coworkers a shared platform with which to collaborate and build on one another's knowledge.

The second lesson of smart swarms is that, as members of such groups, we don't have to surrender our individuality. In nature, good decision-making comes from competition as much as from compromise, from disagreement as much as from consensus. Consider the furious debate that honeybee scouts engage in as they vie for attention for their preferred nest sites. The same is true of human groups: we add something of value to a team or organization mainly by bringing something authentic and original to the table, something that springs from our unique experiences and skills—not by blindly copying others, taking advantage of others, or ignoring our better instincts. At times this means paying our fair share, sacrificing for the good of the group, or accepting the way things are done. At other times it means standing up for what we believe in, lobbying for a cause, or refusing to go along with the crowd. In either case, the best way to serve the group, it turns out, is to be true to ourselves.

In the complicated world of human affairs, after all, you can never be sure how things will work out. Sometimes the audience gets the standing ovation right, and sometimes it doesn't. Sometimes a financial market balances the desires of countless competing interests and sometimes it lets us down. Sometimes it's okay to run your air conditioner on a hot day, and sometimes it helps trigger a blackout. Because we find it so difficult to understand the complex systems we're part of, we might be tempted to give up and simply do what others are doing. But if standing ovations are any guide, you can be sure of one thing: You don't want to go through life applauding events you didn't really enjoy. Nor do you want to end up regretting that you passed up a perfectly good chance to stand up and cheer for something truly wonderful.

The bees wouldn't do that, and neither should you.

Acknowledgments

This book could not have been written without the help of the scientists and business leaders who generously shared their time and ideas with me. Deborah Gordon of Stanford University and Thomas Seeley of Cornell University, in particular, were unfailingly insightful and encouraging from start to finish. I am also deeply grateful to John Miller of Carnegie Mellon University; Andrea Cavagna and Irene Giardina of Italy's National Institute for Condensed Matter Physics; Iain Couzin of Princeton University; Vijay Kumar and Michael Kearns of the University of Pennsylvania; Charles Harper of American Air Liquide; Dennis O'Donoghue of Boeing; Marco Dorigo and Jean-Louis Deneubourg of the Free University of Brussels (French); Frank Bryan of the University of Vermont; Deborah Markowitz, Vermont's secretary of state; Michael Mauboussin of Legg Mason Global Asset Management; and Larry Coffin of Bradford, Vermont. All of them took extra time to carefully explain their fascinating projects and give me the benefit of their invaluable experience.

Among those kind enough to grant me key interviews were Doug Lawson of Southwest Airlines; Scott Turner of the State University of New York, Syracuse; Eric Bonabeau of Icosystem; Bob Wiebe of Boeing; Kirk Visscher of the University of California, Riverside; Karsten Heuer of Canmore, Alberta; Frank Heppner of the University of Rhode Island;

Nigel Franks of the University of Bristol; Richard Hackman of Harvard University; Anita Woolley of Carnegie Mellon University; Danny Grünbaum of the University of Washington; Mike Greene of the University of Colorado, Denver; Patrick Laughlin of the University of Illinois at Urbana-Champaign; Clark Hayes and Santiago Olivares of American Air Liquide; Paul Gillies of Montpelier, Vermont; Duncan Watts of Yahoo! Research; James Surowiecki of *The New Yorker*; Alberto Donati of Rome, Italy; John Sterman of MIT's Sloan School of Management; Janet Mueller, Les Music, and Karen Helmer of Boeing; Thomas Schelling of the University of Maryland; Steve Simpson of the University of Sydney; Peter Spurgin of the Australia Plague Locust Commission; Rupert Soar of Freeform Engineering; Jeff Severts of Best Buy; Massoud Amin of the University of Minnesota; Craig Reynolds of Sony; Stephen Regelous of Massive Software; Stephen Pratt of Arizona State University; Regis Vincent of SRI International; Simon Levin and Daniel Rubenstein of Princeton University; Peyton Young of Oxford University; Dirk Helbing and Anders Johansson of the Swiss Federal Institute of Technology, Zurich; and Keith Still of Crowd Dynamics. Thanks also to Karen Raz of Raz Public Relations, Diane Labelle of American Air Liquide, and Jennifer Koures of NuTech Solutions, for their invaluable assistance in having me talk to their clients.

I am especially grateful to the Santa Fe Institute for inviting me to two workshops: "Decisions 2.0: Distributed Decision Making," in Washington, D.C., and "Collective Decision-Making: From Neurons to Societies" in Santa Fe, New Mexico. Thanks to John Miller, Nigel Franks, and Tom Seeley, in particular, for organizing the Santa Fe event and allowing me to hang out with distinguished scientists.

Under the direction of the American Museum of Natural History, the Southwestern Research Station in Arizona's Chiricahua Mountains has been welcoming naturalists and other visitors for more than five decades. I'd like to thank the station staff for making me feel welcome too. I would also like to thank the staff of Shoals Marine Laboratory for allowing me to spend time at their outstanding facility on Appledore Island off the coast of Maine.

ACKNOWLEDGMENTS

I thank Vijay Kumar, Daniela Rus of MIT, and Stephen Morse of Yale University, for letting me attend the "Third Workshop on Swarming in Natural and Engineering Systems," held on Block Island, New York, and Andrea Bertozzi of the University of California at Los Angeles, for welcoming me and my *National Geographic* colleague Todd James to the workshop "Swarming by Nature and by Design," at UCLA's Institute for Pure and Applied Mathematics.

My thanks to Douglas Brinkley and Mike Dash for letting me quote from their terrific books.

The idea for this book came from my agent, David McCormick, and Jeff Galas, a former editor at Avery. Thank you, David and Jeff, for your confidence and vision. Thanks also to Megan Newman, editorial director at Avery, for supporting the project, and to my talented editor, Rachel Holtzman, for her expert guidance and encouragement.

I am deeply indebted to my research assistant, Michelle Harris, for her keen questions and peerless efforts to protect me from my own mistakes.

Kevin Passino of the Ohio State University was kind enough to read the manuscript-in-progress and offer many helpful suggestions. Julia Parrish of the University of Washington, Ashley Ward of the University of Sydney, Chris Rasmussen of the National Geospatial Intelligence Agency, and Scott Page of the University of Michigan also provided expert comments.

To my editors at *National Geographic*, Chris Johns, Victoria Pope, and Tim Appenzeller, I am extraordinarily grateful for allowing me to pursue this book while on the magazine staff. To Anna Maria Diodori and the rest of the team of *National Geographic* Italy, thanks for welcoming my wife, Priscilla, and me to Rome. To Elizabeth Snodgrass, for her excellent research and collaboration on the original article, "Swarm Theory," in the July 2007 issue of *National Geographic*, and to Philip Ball and Tamás Vicsek for making a convincing case for that article, also many thanks.

I thank my friend and fellow author Dan Buettner, for his inspiring example and positive outlook.

To my parents, Robert and Mary Lou Miller, who gave me endless support and encouragement, as well as perceptive comments and suggestions, I will forever be grateful.

Finally, I cannot thank enough my son Matthew and his wife, Lynn; my son Charley and his wife, Kristen; and Priscilla, my lifelong companion, love, and wife, for believing in me and this book.

NOTES

Foreword

Page x: Peter Drucker's description of managers is from *The New Realities in Government and Politics / in Economics and Business / in Society and World View* (New York: Harper & Row, 1989), p. 209.

Page xi: Thomas Stewart cites programmer Craig Reynolds, who used flock rules to create the special effects in *Batman Returns*. See Stewart's *Intellectual Capital* (New York: Doubleday Business, 1997).

Introduction

Page xv: Doug Lawson's virtual ants are a good example of an agent-based model, a type of computer simulation in which individual units, or agents, interact with one another following simple rules of behavior—in this case, choosing a seat on an airplane. Lawson's most recent project at Southwest Airlines is to help design the "lobby of the future" for airport terminals by imagining that space as an ecosystem in which agents such as ticket kiosks or bag drop-off positions compete to serve customers. Those that serve the most customers are allowed to reproduce, like living creatures; those that serve the least die. By feeding his model with real data from Southwest customers, Lawson in effect allows the lobby to design itself through a kind of simulated natural selection.

Page xvii: For more on leaf-cutter ants, see Bert Hölldobler and Edward O. Wilson, *The Superorganism: The Beauty, Elegance, and Strangeness of Insect Societies* (New York: W. W. Norton, 2009), pp. 430–438.

1. Ants

Page 2: My description of life inside Colony 550 is based on Deborah Gordon's decades of research in the New Mexico desert. Every detail in this account comes from her painstaking experiments. For example, it took two summers of experiments by Gordon and her team to confirm the simple fact that harvester ants search for seeds scattered by wind and rain, rather than from local plants. All the other facts about red harvester ants in this chapter are drawn from Gordon's research, including the basic division of their colonies into task groups such as patrollers, maintenance workers, midden workers, and foragers, which was the subject of Gordon's early work with ants.

Page 6: This research by Michael J. Greene and Deborah M. Gordon was published as "Interaction Rate Informs Harvester Ant Task Decisions," in *Behavioral Ecology*, March/April 2007, pp. 451–455.

Page 9: For more on the principle of self-organization, see Scott Camazine, Jean-Louis Deneubourg, Nigel R. Franks, James Sneyd, Guy Theraulaz, and Eric Bonabeau, *Self-Organization in Biological Systems* (Princeton, NJ: Princeton University Press, 2001).

Page 13: Deneubourg and his collaborators published several papers on what became known as the "double bridge" experiments. For two of the more significant ones, see S. Goss, S. Aron, J.-L. Deneubourg, and J. M. Pasteels, "Self-Organized Shortcuts in the Argentine Ant," *Natur-*

wissenschaften 76 (1989), pp. 579–581; and J.-L. Deneubourg, S. Aron, S. Goss, and J.-M. Pasteels, "The Self-Organizing Exploratory Patterns of the Argentine Ant," *Journal of Insect Behavior* 3 (1990), pp. 159–168.

Page 18: For technical details on the ant colony algorithm as an approach to solving the traveling salesman problem, see Marco Dorigo and Thomas Stutzle, *Ant Colony Optimization* (Cambridge, MA: The MIT Press, 2004), pp. 65–119.

Page 18: Dorigo, Maniezzo, and Colorni published their findings in "The Ant System: Optimization by a Colony of Cooperating Agents," in *IEEE Transactions on Systems, Man, and Cybernetics, Part B*, 26, no. 1 (1996).

Page 19: For more on British Telecom's research, see Eric Bonabeau, Marco Dorigo, and Guy Theraulaz, *Swarm Intelligence: From Natural to Artificial Systems* (Santa Fe, NM: Santa Fe Institute, 1999), pp. 85–93.

Page 26: My description of Arthur Samuel's checkers playing program is based on John Holland's extraordinary book, *Emergence: From Chaos to Order* (New York: Basic Books, 1998), pp. 16–19, 143–154.

2. Honeybees

Page 35: Martin Lindauer writes about his observation of scout bees in his book *Communication Among Social Bees* (Cambridge, MA: Harvard University Press, 1967). Tom Seeley describes Lindauer's work in his wonderful book *The Wisdom of the Hive: The Social Physiology of Honey Bee Colonies* (Cambridge, MA: Harvard University Press, 1996).

Page 38: For more on the honeybee house-hunting experiments, see Thomas D. Seeley, P. Kirk Visscher, and Kevin M. Passino, "Group Decision Making in Honey Bee Swarms," *American Scientist* 94 (2006), p. 222.

Page 38: The mechanisms at work in the honeybee swarm's decision making are described in Kevin Passino, Thomas D. Seeley, and P. Kirk Visscher, "Swarm Cognition in Honey Bees," *Behavioral Ecology and Sociobiology* 62, no. 3 (January 2008), p. 404.

Page 44: Jeff Severts's discussion of the gift card sales survey may be found in Renée Dye, "The Promise of Prediction Markets: A Roundtable," *McKinsey Quarterly*, April 2008, pp. 86–87.

Page 44: For more on the insight of nonexperts, see James Surowiecki, *The Wisdom of Crowds* (New York: Anchor, 2004).

Page 48: Scott Page's findings may be found in *The Difference: How the Power of Diversity Creates Better Groups, Firms, Schools, and Societies* (Princeton, NJ: Princeton University Press, 2007).

Page 53: For more on the mock-terrorist-attack experiments, see Anita W. Woolley, Margaret Gerbasi, Christopher F. Chabris, Stephen M. Kosslyn, and J. Richard Hackman, "What Does It Take to Figure Out What Is Going On? How Team Composition and Work Strategy Jointly Shape Analytic Effectiveness," in *The Group Brain Project*, Technical Report No. 4, January 2007. Hackman also discussed these experiments in a talk at the MIT Center for Collective Intelligence on March 5, 2007.

Pages 56–57: For more on the anchoring, status quo, and sunk-cost traps, see John S. Hammond, Ralph L. Keeney, Howard Raiffa, "The Hidden Traps in Decision Making," *Harvard Business Review*, September/October 1998.

Page 58: Eric Bonabeau writes about caveman brain habits in two articles: "When Intuition Is Not Enough: Strategy in the Age of Volatility," *Perspectives on Business Innovation* 9 (2003), pp. 41–47; and "Don't Trust Your Gut," *Harvard Business Review*, May 2003.

Page 59: For more on group decision-making flaws, see Craig D. Parks and Lawrence J. Sanna, *Group Performance and Interaction* (New York: Westview, 1999); and Norbert L. Kerr and R. Scott Tindale, "Group Performance and Decision Making," *Annual Review of Psychology*, 2004.

Page 62: For more on the honeybee swarm as a brain, see Passino, Seeley, and Visscher, "Swarm Cognition in Honey Bees," p. 407.

Page 63: Excellent analyses of the Beer Game may be found in Peter Senge, *The Fifth Discipline: The Art and Practice of the Learning Organization* (New York: Doubleday, 1990), pp. 27–54; and John

Sterman, *Business Dynamics: Systems Thinking and Modeling for a Complex World* (New York: McGraw-Hill, 2000), pp. 684–695.

Page 81: For more on these experiments, see Patrick R. Laughlin et al., "Groups Perform Better Than the Best Individuals on Letters-to-Numbers Problems: Effects of Group Size," *Journal of Personality and Social Psychology* 90, no. 4 (2006), p. 646.

Page 94: Frank Bryan writes about town meetings with great insight and warmth in *Real Democracy: The New England Town Meeting and How It Works* (Chicago: University of Chicago Press, 2004).

Page 102: Robert Putnam's original article "Bowling Alone: America's Declining Social Capital" appeared in *Journal of Democracy* 6, no. 1 (January 1995). He later developed these ideas in his book *Bowling Alone: The Collapse and Revival of American Community* (New York: Simon & Schuster, 2000).

3. Termites

Page 114: Henry Smeathman's report is from "Some Accounts of the Termites Which Are Found in African and Other Hot Climates" (1781), in Erich Hoyt and Ted Schultz, eds., *Insect Lives: Stories of Mystery and Romance from a Hidden World* (New York: John Wiley & Sons, 1999), p. 160.

Page 127: For more on termite nests as homeostasis-regulating structures, see J. Scott Turner, *The Extended Organism: The Physiology of Animal-Built Structures* (Cambridge, MA: Harvard University Press, 2000), and "A Superorganism's Fuzzy Boundaries," *Natural History* 111, no. 6 (July/August 2002).

Page 128: My description of the plane crash that killed Cory Lidle and Tyler Stanger is based on information from the National Transportation Safety Board, Accident Number DCA07MA003, and news accounts such as James Barron, "Manhattan Plane Crash Kills Yankee Pitcher," *The New York Times*, October 12, 2006; Michelle O'Donnell, "Sifting Through the Ruins to Understand a Disaster," *The New York Times*, October 17, 2006; and Carrie Melago and Corky Siemaszko, "Yank Killed in Plane Horror," New York *Daily News*, October 12, 2006.

Page 131: For more on Andrus's seminal paper, see D. Calvin Andrus, "The Wiki and the Blog: Toward a Complex Adaptive Intelligence Community," September 10, 2005.

Page 131: Sean Dennehy and Don Burke talked about the origins of Intellipedia at the Enterprise 2.0 Conference, June 9–12, 2008, in Boston; a video is available online: http://www.e2conf.com/archive/videos/playvideo/index.php?id=641.

Page 132: For more on Wikipedia, see Clay Shirky, *Here Comes Everybody: The Power of Organizing Without Organizations* (New York: The Penguin Press, 2008), pp. 109–142.

Page 139: Michael Kearns published the results of this research in "An Experimental Study of the Coloring Problem on Human Subject Network," *Science* 11 (August 2006), p. 824. For details of his experiments on consensus problems, see Michael Kearns, Stephen Judd, Jinsong Tan, and Jennifer Wortman, "Behavioral Experiments on Biased Voting in Networks," *PNAS* (*Proceedings of the National Academy of Sciences*), February 3, 2009, pp. 1347–1352.

Page 142: For more on "small-world" networks, see Duncan Watts, *Six Degrees: The Science of a Connected Age* (New York: W. W. Norton, 2003), pp. 69–100; and Duncan Watts and Steven Strogatz, "Collective Dynamics of 'Small-World' Networks," *Nature* 4 (June 1998), pp. 440–442.

Page 144: For more on scale-free networks, see Albert-László Barabási, *Linked: How Everything Is Connected to Everything Else and What It Means for Business, Science, and Everyday Life* (New York: Plume, 2003); and Albert-László Barabási and Eric Bonabeau, "Scale-Free Networks," *Scientific American*, May 2003, pp. 50–59.

Page 145: Malcolm Gladwell's book is *The Tipping Point: How Little Things Can Make a Big Difference* (New York: Little, Brown, 2000), pp. 38–59.

Page 150: On Katrina, see Tricia Wachtendorf and James M. Kendra, "Improvising Disaster in the City of Jazz: Organizational Response to Hurricane Katrina," June 2006, at http://understandingkatrina.ssrc.org/Wachtendorf_Kendra.

Page 151: Haley Barbour's remarks appeared in a report by the United States Senate, *Hurricane Katrina:*

A Nation Still Unprepared: Special Report of the Committee on Homeland Security and Governmental Affairs, 2006.

Page 155: The story of the Cajun Navy was vividly told by Douglas Brinkley in *The Great Deluge: Hurricane Katrina, New Orleans, and the Mississippi Gulf Coast* (New York: William Morrow, 2006), pp. 371–381. For more, see Jefferson Hennessy, "The Cajun Navy," *Acadiana Profile*, January/February 2007.

4. Birds of a Feather

Page 159: For more on Edmund Selous's observations, see his *Thought-Transference (or What?) in Birds* (New York: Richard R. Smith, 1931).

Page 161: For more on Selous and collective intelligence, see Iain Couzin's essay "Collective Minds," *Nature*, February 15, 2007.

Page 164: Andrea Cavagna and his colleagues describe their research in Andrea Cavagna et al., "The STARFLAG Handbook on Collective Animal Behaviour: 1. Empirical Methods," *Animal Behaviour* 76, no. 1 (2008), pp. 217–236. Cavagna and Irene Giardina discuss more details in "The Seventh Starling," *Significance*, June 2008.

Page 168: Two earlier studies of flocking were those by Peter F. Major and Lawrence M. Dill, "The Three-Dimensional Structure of Airborne Bird Flocks," *Behavioral Ecology and Sociobiology* 4 (1978), pp. 111–122; and Richard Budgey, "The Three Dimensional Structure of Bird Flocks and Its Implications for Birdstrike Tolerance in Aircraft," *International Bird Strike Committee Proceedings* 24 (1998), pp. 307–320.

Page 175: On the Hungarian physicist's model, see Tamás Vicsek et al., "Novel Types of Phase Transition in a System of Self-Driven Particles," *Physical Review Letters*, August 7, 1995, pp. 1226–1229.

Page 183: Massive Software's entertaining website may be visited at www.massive.com. The range of commercials, movies, and other products influenced by the company's creativity is truly impressive. See the article "Model Behaviour," *The Economist*, March 9, 2009, p. 32.

Page 196: For more on Radakov's work, see D. V. Radakov, *Schooling in the Ecology of Fish* (New York: Halsted, 1973).

Page 198: Iain Couzin's schooling models are described in Couzin et al., "Collective Memory and Spatial Sorting in Animal Groups," *Journal of Theoretical Biology* 218 (2002), pp. 1–11.

Page 205: On sticklebacks, schools, and decision making, see Ashley J. W. Ward et al., "Quorum Decision-Making Facilitates Information Transfer in Fish Shoals," *PNAS* (*Proceedings of the National Academy of Sciences*) 105, no. 19, p. 6952; and David J. T. Sumpter et al., "Consensus Decision Making by Fish," *Current Biology* 18 (November 25, 2008), pp. 1773–1777.

Page 212: The experiments in human herding are described by John R. G. Dyer et al., "Leadership, Consensus Decision Making and Collective Behaviour in Humans," *Philosophical Transactions of the Royal Society B (Biological Sciences)* 364, no. 1518 (March 2009), pp. 781–789.

Page 214: For Christakis and Fowler's findings, see Nicholas A. Christakis, M.D., Ph.D., and James H. Fowler, Ph.D., *Connected: The Surprising Power of Our Social Networks and How They Shape Our Lives* (New York: Little, Brown, 2009). See also Christakis and Fowler, "The Spread of Obesity in a Large Social Network over 32 Years," *The New England Journal of Medicine*, July 26, 2007; Fowler and Christakis, "Dynamic Spread of Happiness in a Large Social Network: Longitudinal Analysis over 20 Years in the Framingham Heart Study," *British Medical Journal*, December 4, 2008; and Christakis and Fowler, "The Collective Dynamics of Smoking in a Large Social Network," *The New England Journal of Medicine*, May 22, 2008.

Page 216: For more on the "music market" experiments, see Matthew J. Salganik, Peter Sheridan Dodds, and Duncan J. Watts, "Experimental Study of Inequality and Unpredictability in an Artificial Cultural Market," *Science*, February 10, 2006.

Page 221: On Karsten Heuer, see his *Being Caribou* (Seattle: The Mountaineers Books, 2005). Leanne

Allison's award-winning film may be viewed at http://www.beingcaribou.com/necessaryjourneys/film.html.

5. Locusts

Page 230: On the army ants caught in circling behaviors, see T. C. Schneirla, "A Unique Case of Circular Milling in Ants, Considered in Relation to Trail Following and the General Problem of Orientation," *American Museum Novitates*, no. 1253 (April 8, 1944). William Beebe gives a similar account of milling army ants in his book *Edge of the Jungle* (New York: Henry Holt, 1921).

Page 235: For more on the sensitivity of locusts to touch, see S. J. Simpson, E. Despland, B. F. Hägele, and T. Dodgson, "Gregarious Behavior in Desert Locusts Is Evoked by Touching Their Back Legs," *PNAS* (*Proceedings of the National Academy of Sciences*) 98, no. 7 (March 27, 2001), pp. 3895–3897. For more on the role of serotonin in swarming, see M. L. Anstey, S. M. Rogers, S. J. Simpson, S. M. Rogers, S. R. Ott, and M. Burrows, "Serotonin Mediates Behavioral Gregarization Underlying Swarm Formation in Desert Locusts," *Science*, January 30, 2009, pp. 627–630.

Page 237: For the research by Jerome Buhl and colleagues, see Jerome Buhl et al., "From Disorder to Order in Marching Locusts," *Science*, June 2, 2006, pp. 1402–1406.

Page 243: For more details of the video analysis of the Mina Bridge incident, see Anders Johansson, Dirk Helbing, Habib Z. Al-Abideen, and Salim Al-Bosta, "From Crowd Dynamics to Crowd Safety," *Advances in Complex Systems*, April 2008; and Anders Johansson and Dirk Helbing, "From Crowd Dynamics to Crowd Safety: A Video-Based Analysis," *Advances in Complex Systems*, October 2008.

Page 244: The expert who described shock waves that lifted people out of their shoes was John J. Fruin, in R. A. Smith and J. F. Dickie, eds., *Engineering for Crowd Safety* (Amsterdam: Elsevier Science, 1993), p. 99.

Page 253: On the tulip craze, see Mike Dash, *Tulipomania: The Story of the World's Most Coveted Flower and the Extraordinary Passions It Aroused* (New York: Three Rivers Press, 1999).

Conclusion

Page 259: Thomas Schelling's book *Micromotives and Macroeconomics* (New York: W. W. Norton, 1978) is a classic, combining game theory and economic modeling in a highly approachable narrative for the nonspecialist.

Page 260: For more on the standing-ovation problem, see John H. Miller and Scott E. Page, *Complex Adaptive Systems: An Introduction to Computational Models of Social Life* (Princeton, NJ: Princeton University Press, 2007).

Index

adaptive mimicking
  basic mechanisms of, 163, 223
  benefits of group membership, 204, 224–26
  Candid Camera episode, 209–10
  communication of information, 178–79, 223, 263
  coordination of behavior, 183, 223, 263
  copying, 163, 179, 211–12
  definition and description of, 162–63
  financial investment decisions, 219–21, 254–55
  herd protection, 200, 221–23
  human tendency to copy, 200, 211–12
  nearest-neighbor tracking, 172–78, 189–90, 199, 204, 223, 263
  risk minimization, 199–200, 212–13
  small number of informed individuals, 213
  in social networks, 214–18
  standing ovations, 260–62, 264
  suspension of personal judgment, 220–21, 256, 264
Allison, Leanne, 221, 223–24
American Air Liquide, 20–26, 30, 263
Amin, Massoud, 112–13
Andrus, D. Calvin, 131, 135
Anstey, Mike, 236
ants
  aberrant crowd behavior, 229–31
  adjustments to environmental changes, 4–5, 19–20, 30, 262–63
  antennae contact, 4, 7–9
  carrying of objects, 194–95
  competitive strategies, 194
  computer models based on, xiv–xv, xvii, 17–20, 22–29
  emergence of strategy, 29–30
  local knowledge, 10, 15
  pheromone trails, 13–14, 230
  self-organization, 10, 14, 30, 263
  social structure, xvii–xviii
  task allocation, 1–3, 6–9
Ants at Work (Gordon), 3, 194
Aristotle, 51

Barabási, Albert-László, 145
Beer Game, 64, 66–72
bees
  competition of ideas, 38–39, 263
  cooperation in single goal, 61–63
  decision-making process, 33–35, 63, 263, 264–65
  diversity of knowledge, 39–43, 52, 263
  emergence of decision, 39, 42–43
  narrowing of choices, 43, 263
  waggle dance, 34, 36–38, 40–41, 47–48
  worker piping, 41–42
Being Caribou (Heuer), 222–23
Best Buy, 44, 45–46
Bios Group, 22–25
birds
  collective vigilance, 200
  discernment of numbers of objects, 176–77
  documentation of, 165, 167–71
  flocking behaviors, 159–61, 165–66, 171–72
  nearest-neighbor tracking, 172–78, 263
  simulation models of, 164, 174–75

INDEX

Boeing aircraft company
  diversity of knowledge, 65–66
  production problem, 64–66, 71–72
  social structure, 72–74
  Test Operations Center (TOC), 76–82
  work flow analysis, 74–76
Bonabeau, Eric, 23, 58
Brinkley, Douglas, 155
Bryan, Frank, 94–96, 98–99
Buhl, Jerome, 237
Burke, Don, 131, 137
Burrows, Malcolm, 236

Candid Camera (television program), 209–10
caribou herd, 221–23
cascade of events
  definition of, 111
  financial crisis, 251
  in highly connected networks, 111
  in social networks, 215
  standing ovations, 262, 264
Cavagna, Andrea. *See* Starlings in Flight (STARFLAG) project
caveman brains
  counterterrorism experiment, 53–55, 60
  groups, 59–60
  individuals, 56–59
  parallel processing of information, 62–63
  reductive bias, 71
checker-playing model, 26–28
Christakis, Nicholas, 214–15
Coffin, Larry, 86–91, 94
collaboration. *See* indirect collaboration
collective vigilance, 200
Colorni, Alberto, 16
combinatorial optimization problems, 16–19
communication. *See also* indirect collaboration
  adaptive mimicking and nearest-neighbor tracking, 178–79, 189–90, 199, 204, 223, 263
  ant antennae-touching, 4, 7–9
  bee waggle-dancing, 34, 36–37
  bee worker-piping, 41–42
  decentralized, 152–54
  lack of, within crowd, 247
  through pheromones, 13–14, 17–19, 24–25, 123, 125, 230
  among robots, 188–89
competition of ideas
  bees, 38–39, 263
  for effective decision making, 43, 52, 263
  town meetings, 84, 93, 95
Complex Adaptive Systems (Miller and Page), 260–61, 262
coordination of behavior, 183, 223, 263
copying. *See* adaptive mimicking
counterterrorism experiment, 53–55, 60
Couzin, Iain
  fish-schooling research, 198–99, 200–202, 204–9
  on locust cannibalism, 255
  on propagation of information within groups, 213
Crawley, Jon, 126
crowds
  aberrant behavior, 229–31, 233–34, 238–39, 246–47, 255
  breakdown of individual control, 246–47, 264
  computer simulation of, 180–85
  control of, 247–48, 256
  emergent behavior, 170, 182–83
  financial market behavior, 249–55
  flow patterns and shock waves, 243–45
  herding behavior, 213
  lack of communication within, 247
  self-organization, 246
  tramplings, 227–29, 239–42
  wisdom of, 44–52

Dash, Mike, 253–54
decentralized behavior
  ant self-organization, 10–11, 17, 19
  bee nest selection, 39, 63
  human communication channels, 152–57
Delius, Juan D., 176
Deneubourg, Jean-Louis, 13–15, 121
Dennehy, Sean, 131–32, 134, 136
Deutsch, Andy, 67
Difference, The (Page), 49
distributed collaboration
  Intellipedia, 130–32, 134–36, 137–38
  Wikipedia, 130, 132–34
  wikis, 130–32, 134–35, 153–54
distributed problem-solving
  collection of unique information by individuals, 39
  Optimal Coloring Problem, 141
  self-healing system, 112, 125, 154
  self-organizing system, 10–11, 17, 19–20
distributed robotics, 189–90
diversity of knowledge
  bee decision making, 39–43, 52, 63, 263
  Beer Game, 64, 66–72
  biases and bad habits, 56–60

Boeing aircraft company, 65–66, 77
cognitive diversity, 49, 54
combining of information, 49–51
counterterrorism experiment, 53–55, 60
definition and description of, 39
emergence of behavior, 39, 42–43, 92
prediction markets, 45–47
small groups, 81–82
for successful decision making, 48–52
town meetings, 84, 93–94, 97–99
wisdom of crowds, 44–52, 263
Donati, Alberto, 23
Dorigo, Marco, 12–13, 15–19, 190–92

Electric Power Research Institute (EPRI), 111–12, 125
Emergence (Holland), 28
emergence of behaviors
in crowds, 170, 182–83
from local interactions, 197
in self-organizing system, 29–30, 135
through diversity of knowledge, 39, 42–43, 92
Emmerton, Jacky, 176
Extended Organism, The (Turner), 125–26

Fingar, Thomas, 134–35
fish
emergence of behavior, 197
nearest-neighbor tracking, 199, 204
phase transition, 202–3
schooling patterns, 201–4
threshold response, 205–9
waves of agitation, 197–98
flocking behavior. *See* birds
Fowler, James, 214–15
Friedman, Thomas, 252

Giabosi, Margaret, 53
Gillies, Paul, 96
Gladwell, Malcolm, 145–46
Gordon, Deborah, 1, 3, 5–9, 10, 31–32, 194
Grassé, Pierre-Paul, 120
Great Deluge, The (Brinkley), 155
greedy function, 24
Greene, Mike, 2–3, 7–8, 12
groupthink, 60
Guare, John, 143

Hackman, Richard, 53–56, 60, 82
hajj
flow patterns and shock waves, 242–45
rituals, 239–40
safety strategies, 240–41, 247–48
signs of impending trouble, 245–46
tramplings, 240, 241–42
Hammond, John, 57
Hanson Robotics, 185
Harper, Charles, 20–23, 26
Harter, Pascale, 232
Hayes, Clarke, 22
Helbing, Dirk, 242–45
Helmer, Karen, 75
Heppner, Frank, 162, 166, 168–69, 225
herding behavior. *See* adaptive mimicking
Here Comes Everybody (Shirky), 134
Heuer, Karsten, 221–26
Hewlett-Packard lab (England), 19–20
"Hidden Traps in Decision Making, The" (Hammond, Keeney, and Raiffa), 57
Holland, John, 28–29
Holland tulip craze, 252–54
Hölldobler, Bert, xvii–xviii
honeybees. *See* bees
humans. *See also specific behaviors*
imbalance between individual and collective interests, 265–66
lessons learned from swarms, 263, 267–69
sense of belonging within herd, 224–26
tendency toward copying, 200, 211–12
Hurricane Katrina, 150–57

Icelandic financial crisis, 249–52, 256–57
imitation. *See* adaptive mimicking
indirect collaboration
decentralized disaster response, 152–54
definition and description of, 121–22
Intellipedia, 130–31, 132, 138
termite mound construction, 120–22, 263
Web 2.0 tools for, 136–37
Wikipedia, 130, 132–34
wikis, 130–32, 134–35, 153–54
Intellipedia, 130–32, 134–36, 137–38
Ioannou, Christos, 203–4

Jackson, Peter, 180–81
James, William, 161
Janis, Irving, 60
Jeremica, Vern, 67

Kearns, Michael, 140–42, 144, 146–49
Keeney, Ralph, 57
Kendra, James M., 154
Kosslyn, Stephen, 53
Kumar, Vijay, 193–96

INDEX

Laughlin, Patrick, 81–82
Lawson, Doug, xiv–xvi, xvii
Leonard, Naomi, 198–99
Levin, Simon, 211
Lindauer, Martin, 35–37
Linked (Barabási), 145
locusts
  behavioral and physical transformations, 232–34
  cannibalism, 238–39, 255
  overcrowding, 237–38
  rear-leg sensitivity, 235–36, 239
  serotonin release, 236
  2004 swarms, 231–32, 233
Lord of the Rings (movies), 179–83

Mackay, Charles, 255
Maniezzo, Vittorio, 17–18
Markowitz, Deborah, 100–102, 103
Massive Software, 183–86
Mauboussin, Michael, 220–21
Mecca pilgrimage. *See* hajj
Memoirs of Extraordinary Popular Delusions (Mackay), 255
Micromotives and Macrobehavior (Schelling), 266
Miller, John, 260–61, 262, 264
Minority Report (movie), 186–87
More Than You Know (Mauboussin), 220–21
Mueller, Janet, 78

narrowing of choices, 43, 84, 93, 95–97, 263
nearest-neighbor tracking, 172–78, 189–90, 199, 204, 223, 263
networks. *See also* social networks
  power grid, 106–12, 144
  scale-free, 144–46
  small-world, 142–44
New York
  airplane crash, 128–30, 138–39
  2003 blackout, 107–8
NuTech Solutions (formerly Bios Group), 22–25

O'Donoghue, Dennis, 63–68, 70–77, 80–81, 149
Optimal Coloring Problem
  distributed problem-solving, 139–41
  effect of network structure, 146–49
  network types, 142–46
O'Reilly, Tim, 136
Ortiz, Charles, 189
Overbye, Thomas, 110

Page, Scott, 48–52, 60–61, 94, 260–61, 262
Parisi, Giorgio, 164
Passino, Kevin, 61–63
phase transition, 202–3
pheromones
  ant trails, 13–14, 230
  in termite mound construction, 123, 125
  virtual, in ant-based computer models, 17–19, 24–25
Politics (Aristotle), 51
Pomeroy, Harold, 168–69
power grid
  chain of failures, 106–9
  complexity and interconnectedness, 110–11
  self-healing, 111–12
  as small-world network, 144
Pratt, Stephen, 193–95
prediction markets, 45–47
Putnam, Robert, 102–3

Radakov, Dmitri, 196–98, 212
Raiffa, Howard, 57
red harvester ants. *See* ants
reductive bias, 71
Regelous, Stephen, 179–86
Reynolds, Craig, 174–75
Robert, Leon, III, 67
Roberts, Sara, 155–56
Robert's Rules of Order, 93–94, 95–96
robotic swarms, 186–93
Rubenstein, Daniel, 211

Samuel, Arthur, 26–29
scale-free networks, 144–46
Schelling, Thomas, 259–60, 265–66
Schewe, Philip, 112–13
Schneirla, T. C., 229–31
schooling behavior. *See* fish
Seeley, Thomas, 33–34, 37–42, 47–48, 84, 264–65, 267
self-healing systems
  power grid, 111–12
  in smart swarms, 111
  termite mounds, 122–27
self-organization
  ants, 10, 14, 30, 263
  basic mechanisms of, 10–12
  blogs, 131, 135
  computer models of, 15–20, 22–26
  crowds, 246
  definition and description of, 9–10
  emergence of strategy, 29–30, 135
  flexibility in response to environmental changes, 19–20, 30
  greedy function, 24

intelligence-gathering, 131, 135
peer groups, 211
robots, 191–92
stigmergy process, 121
termites, 121
wikis, 130–32, 134–35, 153–54
Selous, Edmund, 159–62
serotonin, 236
Severts, Jeff, 44, 45–46, 59
Shirky, Clay, 134, 137
Sidgwick, Henry, 161
Simpson, Steve, 234–36, 239
Six Degrees (Watts), 11
Six Degrees of Separation (Guare), 143
small-world networks, 142–44
smart-swarm principles. *See also individual principles*
adaptive mimicking, 162–63
diversity of knowledge, 39
indirect collaboration, 121–22
self-organization, 9–10
smart swarms
definition of, xvii
lessons learned from, 263, 267–69
Smeathman, Henry, 114
Soar, Rupert, 116–19, 126
social networks
disaster response, 152–54
effect of structure on outcome, 141–50
Intellipedia, 130–31, 132, 138
product marketing in, 218–21
spread of influence throughout, 145–46, 214–18
Web 2.0 tools for, 136–37
Wikipedia, 130, 132–34
wikis, 130–32, 134–35, 153–54
Southwest Airlines, xiii–xvi
Spurgin, Peter, 233–34
SRI International, 188
standing ovations, 260–62, 264
Starlings in Flight (STARFLAG) project
distribution of birds in flock, 171–72
goal, 164–67
methodology, 167–68, 170–71
nearest-neighbor tracking, 172–78
Stephenson, David, 154–55
stigmergy, 120–21, 133, 138
Still, Keith, 245–46, 248, 256
Sumpter, David, 204–6
Superorganism, The (Hölldobler and Wilson), xvii–xviii
Surowiecki, James, 44–45, 48, 52, 255, 256
Swarm-bots project, 190–92, 196
Tapscott, Don, ix–xii, 137
termite mounds
balancing of environmental factors, 117–18, 124–25, 127–28
damage repair, 119–20, 122–25, 126–27
digital model of, 116–17
function and structure, 113–17
indirect collaboration in construction process, 120–22, 263
threshold response, 205–9, 237–38
Tipping Point, The (Gladwell), 145–46
town meetings. *See* Vermont town meetings
traveling salesman problem, 16–19
Tulipomania (Dash), 253–54
Turner, J. Scott, 114, 116–18, 119–20, 122–28

Vermont town meetings
compared to bee behavior, 84
competition of ideas, 95
discussion and voting, 87–91
diversity of knowledge, 97–99
empathy for neighbors, 101–3
issues and options, 83, 85–87, 94–95
moderators' roles, 92–93
narrowing of choices, 95–97
Robert's Rules of Order, 93–94, 95–96
Vicsek, Tamás, 175, 202
Vincent, Regis, 189
Visscher, Kirk, 33–34, 37–42
von Frisch, Karl, 36

Wachtendorf, Tricia, 154
Ward, Ashley, 204–6
Watts, Duncan, 111, 212, 216–19
Web 2.0 collaboration, 136–37
Who Wants to Be a Millionaire? (television program), 44–45, 49–50
Wiebe, Bob, 75
Wikinomics (Tapscott and Williams), 137
Wikipedia, 130, 132–34
wikis, 130–32, 134–35, 153–54
Williams, Anthony D., 137
Wilson, E. O., xvii–xviii
wind energy, 118–19
wisdom of crowds, 44–52, 263. *See also* diversity of knowledge
Wisdom of Crowds, The (Surowiecki), 44–45
Woolley, Anita, 53–55, 60, 82

Young, Peyton, 219–20